T0189375

Advances in Intelligent Systems and Computing

Volume 864

Series editor

Janusz Kacprzyk, Polish Academy of Sciences, Warsaw, Poland
e-mail: kacprzyk@ibspan.waw.pl

The series "Advances in Intelligent Systems and Computing" contains publications on theory, applications, and design methods of Intelligent Systems and Intelligent Computing. Virtually all disciplines such as engineering, natural sciences, computer and information science, ICT, economics, business, e-commerce, environment, healthcare, life science are covered. The list of topics spans all the areas of modern intelligent systems and computing such as: computational intelligence, soft computing including neural networks, fuzzy systems, evolutionary computing and the fusion of these paradigms, social intelligence, ambient intelligence, computational neuroscience, artificial life, virtual worlds and society, cognitive science and systems, Perception and Vision, DNA and immune based systems, self-organizing and adaptive systems, e-Learning and teaching, human-centered and human-centric computing, recommender systems, intelligent control, robotics and mechatronics including human-machine teaming, knowledge-based paradigms, learning paradigms, machine ethics, intelligent data analysis, knowledge management, intelligent agents, intelligent decision making and support, intelligent network security, trust management, interactive entertainment, Web intelligence and multimedia.

The publications within "Advances in Intelligent Systems and Computing" are primarily proceedings of important conferences, symposia and congresses. They cover significant recent developments in the field, both of a foundational and applicable character. An important characteristic feature of the series is the short publication time and world-wide distribution. This permits a rapid and broad dissemination of research results.

More information about this series at http://www.springer.com/series/11156

Bernadetta Kwintiana Ane
Andi Cakravastia · Lucia Diawati
Editors

Proceedings of the 18th Online World Conference on Soft Computing in Industrial Applications (WSC18)

 Springer

Editors
Bernadetta Kwintiana Ane
Institute of Computer-Aided Product
  Development Systems
University of Stuttgart
Stuttgart, Germany

Lucia Diawati
Industrial Engineering
Bandung Institute of Technology
Bandung, Indonesia

Andi Cakravastia
Department of Industrial Engineering
Bandung Institute of Technology
Bandung, Indonesia

ISSN 2194-5357   ISSN 2194-5365 (electronic)
Advances in Intelligent Systems and Computing
ISBN 978-3-030-00610-5   ISBN 978-3-030-00612-9   (eBook)
https://doi.org/10.1007/978-3-030-00612-9

Library of Congress Control Number: 2018954075

This Springer imprint is published by the registered company Springer Nature Switzerland AG
The registered company address is: Gewerbestrasse 11, 6330 Cham, Switzerland

Preface

Continuing a tradition started over a decade ago by the World Federation of Soft Computing, WSC18 Conference aims at bringing together outstanding research and developments in the field of Soft Computing and its applications in industries from across the world. WSC18 strives to foster an open and lively scientific forum by gathering researchers, engineers, and practitioners with diversified backgrounds to discuss and disseminate the latest state-of-the-art achievements, thought-provoking challenges, and potential failure directions in many different aspects.

This volume of Advances in Intelligent Systems and Computing contains the accepted papers presented in the 18th Online World Conference on Soft Computing in Industrial Applications (WSC18) held on December 1–12, 2014 at the World Wide Web. The conference is open to all academics, students, and industrial/commercial parties, and hosted online between authors and participants over the Internet.

The 2014 edition of the Online World Conference on Soft Computing in Industrial Application consists of selected keynote speech, invited talks, tutorials, special sessions, and general track papers. The program committee received a total of 51 submissions from 12 countries, which reflect the international nature of this event. Each paper was peer-reviewed by typically three referees, culminating in the acceptance of 20 papers for publication. The review process required an enormous effort from the members of the International Technical Program Committee, and we would therefore like to thank all its members for their contribution to the success of this conference. We would like to express our sincere gratitude to the special session organizers, to the host of WSC18, Bandung Institute of Technology in Indonesia, and to the publisher, Springer, for their vigorous work and support in organizing the conference. Our special thanks to Aisha F. Chairunnisa, Daniel P. Sembiring, Ilham M. Lufi, and Ryan A. Moniaga for professional editing works. Finally, we would like to thank all the authors for their high-quality contributions. *"Desire and passion in science and research, and spirit to share it around the globe made this event a success!"*

August 2018   Bernadetta Kwintiana Ane
                                                                             Andi Cakravastia
                                                                             Lucia Diawati

Organization

General Chairs

Andi Cakravastia — Bandung Institute of Technology, Indonesia
Bernadetta Kwintiana Ane — University of Stuttgart, Germany

Program Chairs

Anas Maruf — Bandung Institute of Technology, Indonesia
Fariz Hasby — Bandung Institute of Technology, Indonesia

Special Event Chairs

Vaclav Snasel — VSB-TU Ostrava, Czech Republic
Pavel Kromer — VSB-TU Ostrava, Czech Republic

Publicity Chair

Jude Hemanth — Karunya University, India

International Advisory Board

Kalyanmoy Deb — Indian Institute of Technology Kanpur, India
Carlos M. Fonseca — University of Algarve, Portugal

Thomas Stützle   University Libre de Bruxelles, Belgium
Jörn Mehnen   Cranfield University, UK
Mario Köppen   Kyushu Institute of Technology, Japan
Xiao-Zhi Gao   Aalto University, Finland

International Technical Program Committee

Alexandre Delbem, Brazil
Ana Viana, Portugal
Ana Maria Rocha, Portugal
Anderson Duarte, Brazil
André Carvalho, Brazil
André Coelho, Brazil
André Cruz, Brazil
Bayesian Methods
Andreas Koenig, Germany
Aurora Pozo, Brazil
Bernard Grabot, France
Carlos Henggeler Antunes, Portugal
Carlos Coello Coello, Mexico
Daniel Gonzáles Peña, Spain
Eduardo Carrano, Brazil
Eiji Uchino, Japan
Elizabeth Wanner, Brazil
Felipe Campelo, Brazil
Fengxiang Qiao, USA
Fernanda Costa, Portugal
Fernando Von Zuben, Brazil
Francisco Herrera, Spain
Francisco Pereira, Portugal
Frank Klawonn, Germany
Frederico Guimarães, Brazil
Gina Oliveira, Brazil
Giovanni Semeraro, Italy
Gisele Pappa, Brazil
Guy De Tré, Belgium
Heitor Lopes, Brazil
Helio Barbosa, Brazil
Hisao Ishibuchi, Japan
Ioannis Hatzilygeroudis, Greece
Jae Oh, USA
Janos Abonyi, Hungary
Jerzy Grzymala-Busse, USA

João Vasconcelos, Brazil
Justin Dauwels, USA
Keshav Dahal, UK
Leandro Coelho, Brazil
Leandro de Castro, Brazil
Lino Costa, Portugal
Luís Paquete, Portugal
Luiz Duczmal, Brazil
Marcin Paprzycki, Poland
Marcone Souza, Brazil
Maria Teresa Pereira, Portugal
Michele Ottomanelli, Italy
Olgierd Unold, Poland
Oriane Neto, Brazil
Oscar Castillo, Mexico
Patricia Melin, Mexico
Petra Kersting, Germany
Petrica Pop, Romania
Radu-Emil Precup, Romania
Renato Krohling, Brazil
Roderich Gross, UK
Rodrigo Cardoso, Brazil
Rui Pereira, Portugal
S. G. Ponnambalam, Malaysia
Sanaz Mostaghim, Germany
Santosh Kumar Nanda, India
Sara Silva, Portugal
Sudhirkumar Barai, India
Tobias Wagner, Germany
Thomas Stuetzle, Belgium
Uri Kartoun, USA
Valeriu Beiu, United Arab Emirates
Viviane G. da Fonseca, Portugal
Yos Sunitiyoso, Indonesia
Zvi Boger, Israel

Webmasters

Mugi Sugiarto	Bandung Institute of Technology, Indonesia
Benirio Hermanto	Bandung Institute of Technology, Indonesia
Aris Triyanto	Bandung Institute of Technology, Indonesia

Sponsoring Institutions

World Federation on Soft Computing
Bandung Institute of Technology, Indonesia

Contents

Some Thoughts of Soft Computing

Bio-inspired Algorithms in Application

Three Research Directions of Interactive Evolutionary Computation

Hideyuki Takagi$^{(\boxtimes)}$

Faculty of Design, Kyushu University, Fukuoka, Japan
takagi@design.kyushu-u.ac.jp

Abstract. We overview three research directions of interactive evolutionary computation (IEC). They are (1) extending IEC applications, (2) reducing IEC user fatigue by improving/ developing interface, algorithms, operators, frameworks, and others, and (3) using IEC as a tool for analyzing human characteristics.

1 Introduction

Interactive evolutionary computation (IEC) is a framework that evolutionary computation (EC) optimizes a target system based on human subjective evaluations to the outputs from the target system. There are many tasks which performances are hard to be measured but can be evaluated by humans. IEC shown in Fig. 1 (a) can optimize such tasks by involving a human user in an optimization loop.

2 IEC Application-Oriented Research

The first direction of IEC research is to expand IEC application areas. IEC has been applied to wide variety of areas. They are roughly categorized into three:

1. artistic applications such as creating computer graphics (CG), music, editorial design, and industrial design,
2. engineering applications such as acoustic or image processing, robotics control, data mining, generating programming code, and media database retrieval, and
3. others such as educations, games, and geological simulation.

See these perspectives in the reference [1].

3 Research for Reducing IEC User Fatigue and Making IEC Practical

The second direction of IEC research is to reduce IEC user fatigue and make IEC practical. IEC users must repeat evaluations many times and feedback them to a tireless computer. This nature causes IEC user fatigue, and especially, it becomes a serious problem for practical use when end-users use IEC.

© Springer Nature Switzerland AG 2019
B. K. Ane et al. (Eds.): WSC 2014, AISC 864, pp. 3–5, 2019.
https://doi.org/10.1007/978-3-030-00612-9_1

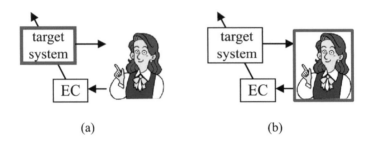

Fig. 1. IEC frameworks (a) for optimization and (b) for human science

Several trials have been proposed. Some of them are: improving IEC user interface, developing new EC algorithms and EC operations that make EC converge faster and are effective under the restricted IEC conditions, developing new IEC framework with less human fatigue is other solution, introducing an IEC user's evaluation model made by machine learning, allowing an IEC user to intervene in an EC search, and others. See these works in the reference [1].

4   IEC as a Tool for Analyzing Human Characteristics

The third direction of IEC research is to use the IEC as a tool for analyzing human characteristics; see Fig. 1 (b). This is a new and unique approach, and there are few related works so far.

Since an IEC target system is optimized based on a human psychological evaluation scale, we may know the scale indirectly by analyzing the optimized target system. It somehow has similarity to *reverse engineering* in software engineering. Other explanation of this approach is that IEC is a tool for visualizing impressions or images in mind. Artists have skills for expressing them by drawing pictures, playing musical instruments, programming CG and writing in poems, for example. However, it is hard for many ordinary people who have no such skills to express the impressions or mental images. IEC helps those who have less skill to express the mental images using IEC-based systems.

Thanks to this kind of IEC use, we may be able to analyze human characteristics by analyzing obtained optimized systems and their system outputs. There are few research along to this research direction. Some of them are measuring mental dynamic range of mental patients [2], finding new facts of an auditory system, modeling of awareness mechanism [3]. Through the analysis, we are looking forward to finding out new psychological or physiological unknown facts.

References

1. Takagi, H.: Interactive evolutionary computation: fusion of the capabilities of EC optimization and human evaluation. Proce. IEEE **89**(9), 1275–1296 (2001)
2. Takagi, H., Takahashi, T., Aoki, K.: Applicability of interactive evolutionary computation to mental health measurement. In: IEEE International Conference on Systems, Man, and Cybernetics (SMC 2004), The Hague, The Netherlands, pp. 5714–5718, 10–13 October 2004
3. Takagi, H.: Interactive evolutionary computation for analyzing human aware mechanism. Appl. Comput. Intell. Soft Comput. **2012** (2012) Article ID 694836, https://doi.org/10.1155/2012/694836

Adaptive Information Processing in Computer-Aided Product Development

Dieter Roller$^{(\boxtimes)}$ and Bernadetta Kwintiana Ane

Institute of Computer-Aided Product Development Systems,
Universität Stuttgart, Stuttgart, Germany
`{roller,ane}@informatik.uni-stuttgart.de`

Abstract. The mechanical and electrical design aspects of mechatronic systems are highly intertwined through a substantial number of constraints existing between their components. A simple change in mechanical CAD may entail a redesign of several parts in electrical CAD and vice versa. Here, parametric constraints are used to model dependencies between the geometry of objects. In this situation, an active semantic network is used that provides support to an active, data-driven design process that controls parallel modifications in the shared database and keeps designers being informed when their work coincides with parallel work. The architecture allows different users or different function-groups and departments within a company to configure their own environment by modifying the information in the user/group model and to add additional filters for adaptive information processing.

Keywords: Parametric constraints · Active semantic network
Adaptive information processing

1 CAD Design Today

Years ago, when feature-based parametric modelling was first released, it really did revolutionise the computer-aided design (CAD) industry. It fundamentally changed the way that engineering organisations developed 3D models, and how they have made changes to the designs as well. Feature-based modelling approach allows designers to use features that correspond to physical entities to construct solid models, instead of dealing with primitive geometric entities, such as points, curves, and solid primitives. Whilst, parametric modelling approach enables the use of parameters and constraints to drive object size and location to produce designs with features that adapt to changes made to other features. By adopting a parametric approach to create models, designers are setting the clear parameters, features and relationships of the models, which is intended to capture the product's behaviour [1]. This modelling approach allows designers to create solid models in such a way that by varying a few parameters (e.g., geometric dimensions), the solid models rebuild automatically as intended by capturing the design intent [2]. It updates parts of the model accordingly, if changes occur to the design and coupled with a complete bi-directional association between parts.

Feature-based parametric modellers today offer the ability to maintain a history of the modelling process, which typically appears in a feature tree or history tree, see

© Springer Nature Switzerland AG 2019
B. K. Ane et al. (Eds.): WSC 2014, AISC 864, pp. 6–21, 2019.
https://doi.org/10.1007/978-3-030-00612-9_2

Fig. 1. History-based parametric models enable to capture design intent effectively and maintain design constraint. However, maintaining detailed design history can add difficulty to the design process, particularly when significant changes are required early in the model history. It increases also complexity in the design process when designers working on a collaborative environment have to share a model with others who do not understand the design history or who use different software.

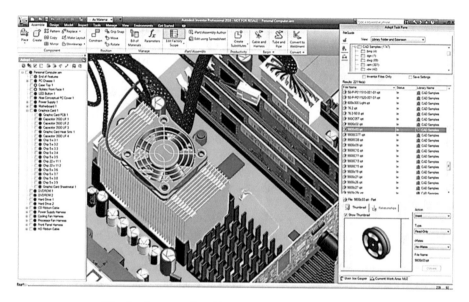

Fig. 1. An example of feature-based parametric model

2 Challenges in Mechatronic System Design

The development of mechatronic products require a wide range of disciplines and hence are collaborative in nature. A mechatronic system encompasses mechanical, electronics, electrical and software components. The design of mechanical components requires a sound understanding of core mechanical engineering subjects, including mechanical devices and engineering mechanics. Electronics involves measurement systems, actuators, and power control. The electrical aspect of mechatronic systems involves the functional design of electrical plants and control units. Often, the collaborative development of mechatronic systems is inefficient and error-prone because contemporary design environments do not allow sufficient flow of design and manufacturing information across electrical and mechanical domains [3].

The amount of engineering related information required for the design of complex mechatronic products tends to be enormous. Aspects to be considered include geometric shapes of mechanical components, electrical wiring information, information about input and output pins of electronic circuit boards, and so on. Today, the integrated design of mechatronic systems can be facilitated through the integration of mechanical and electrical CAD systems. One approach to achieve such integration is through the propagation of constraints [4].

3 MCAD–ECAD Integration

The mechanical and electrical design aspects of mechatronic systems are highly intertwined through a substantial number of constraints existing between their components. Consequently, in order to integrate mechanical and electrical CAD tools on systems integration level into an overarching cross-disciplinary computer-aided engineering (CAE) environment, these constraints have to be identified, understood, modelled, and bi-directionally processed.

Constraints in mechanical domain can be classified into geometric, kinematics, force, energy, and material constraints. In electrical domain, constraints are devised into electrical resistance, electrical capacitance, electrical inductance, motor torque, and system control. Modifications made on mechanical CAD (MCAD) site may lead to significant design modifications to be made on electrical CAD (ECAD) site and *vice versa*. Obviously, there exist a huge number of constraints between a mechanical part of a mechatronic product and its electrical counterpart that have to be fulfilled to have a valid design configuration.

An MCAD model typically contains the following information: *features*, which are high-level geometric constructs used during the design process to create shape configurations in the CAD models; *parameters*, which are values of quantities in the CAD model, such as dimensions; *constraints*, which are relationships between geometric elements in the CAD models, such as parallelism, tangency, and symmetry. An MCAD system cannot simply transfer such information to other CAD/CAE system because these systems have significantly different software architectures and data models.

In order to develop a shared knowledge base for interdisciplinary parametric product modelling, an approach so-called Active Semantic Network (ASN) can be used [5]. The ASN is a shared database management system (DBMS) whose mechanisms have been adapted to the requirements of an active, data-driven design process. Such an ASN can serve as a backbone to MCAD–ECAD integration since it can be used as a common workspace for designers involved in product design. In ASNs, constraints are used to model dependencies between interdisciplinary product models and cooperation, and rules using these constraints are created to help designers to collaborate and integrate their results to a common solution. With this design approach, designers have the ability to visualise the consequences of design decisions across disciplines.

4 Active Semantic Network

A semantic network is a graphic notation for representing knowledge in patterns of interconnected nodes and arcs. It is a graph that consists of vertices, which represent concepts, and arcs, which represent relations between concepts. An ASN can be realised as an active, distributed, and object-oriented DBMS (database management system) [6]. An active DBMS allows users to specify actions to be taken automatically when certain conditions arise.

A database object in the ASN consists of the data itself, a set of associated rules, and cooperative locks. Constraints are modelled as normal database objects that are also subject to user modifications. The active behaviour of the ASN is provided by an active rule component. This component uses event-condition-action (ECA) rules of active database to realise active mechanisms [7]. An ECA rule consists of an event that triggers a rule, a condition and an action. The condition is a collection of queries that is evaluated when a rule is triggered by an event. The action of a rule is executed when the event is triggered and the condition is satisfied. Inside an action, inferences can be drawn as well as external applications can be started like communication or cooperation tools. With ECA rules, a designer can define geometrical constraints in parametric product models as well as complex and interdisciplinary inter-object dependencies. The ECA rules linked to a constraint object specify the active behaviour of the constraint and are evaluated at run time.

Figure 2 describes the corresponding ASN system architecture. The system consists of some advanced database features including a distributed object management, an active rule component, a parametric constraint management, and a cooperative transaction model. There exists constraint propagation within the ASN. The ASN uses a rule-based evaluation method for tracking of constraint propagation. When a constraint is violated, possible actions include extensive inferences on the product data and notifications will be sent to the responsible designers to inform them about the violated constraints.

5 Adaptive Information Processing

In CAD modelling, parametric constraints are used to model dependencies between the geometry of objects. These dependencies are described by mathematical formulas and constraint propagation mechanisms enforce these constraints to be met. Using interdisciplinary parametric constraints (e.g., constraints between MCAD and ECAD models), usually no general constraint satisfaction algorithms can be defined. A simple change in MCAD may entail a redesign of several parts in ECAD and vice versa. In this situation, the ASN provides support to an active, data-driven design process that controls parallel modifications in the shared database and keeps designers being informed when their work coincides with parallel work.

To support cooperative work, instead of locking objects when designers are modifying them and preventing that other users are able to access these objects, the cooperative transaction system informs the designers about conflicts and provides several methods of conflict resolution. When the involved users decide to incorporate, the cooperative transaction system provides group transactions that allow transactions to share data and to exchange information. When designers decide to work independently on different design versions, the active notification mechanism can be used to inform them whenever other users have reached new partial results. Also, read access is enabled to parallel versions in order to allow reading the results of other users and to adapt their work early to these results.

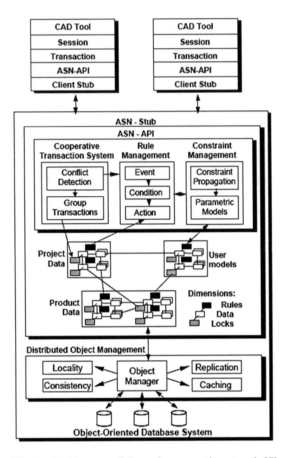

Fig. 2. Architecture of the active semantic network [5]

In this scheme, the cooperative transaction model [8] introduces different forms of concurrency control. In comparison with the traditional database management systems, they are based on the notion of serializability and ACID (atomicity, consistency, isolation, durability) principles, but are not appropriate to deal with the interactive and cooperative nature of current product design. Here instead of guaranteeing data consistency by restricting concurrency control, the transaction model ensures the safety of the design process. Serializability and atomicity are replaced by new informal models based on user notification and new policies in group transactions to provide new notions of correctness in cooperative work.

The adaptive component of the database system provides methods for individual presentation of information with respect to the actual user. Based on the user model mentioned earlier, an intelligent agent determines the presentation form suitable for the specific user.

5.1   Architectural Model and Levels of Adaptive Information Processing

Adaptive information processing interferes with the high-level user interface. This is the module of the information system that users directly interact with and which has to deal with how the users want to use the system. It is the task of this level in combination with the adaptive information processing agent to create a sort of transparency to the underlying database for the information searching person and to enhance the functionality of the intelligent database. Figure 3 shows the underlying architectural model of this concept.

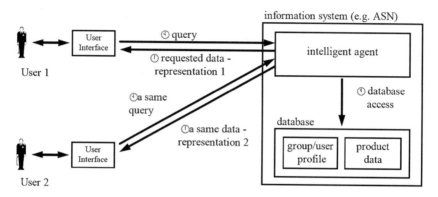

Fig. 3. Architectural model of adaptive client/server-based information processing [9]

The architecture allows different users or different function-groups and departments within a company to configure their own environment by modifying the information in the user/group model and to add additional filters for adaptive information processing.

In general, the adaptive information processing can be implemented at different levels:

- level-0: The intelligent agent chooses a form suitable to present data (e.g., spreadsheet data, a table, some charts, or a company report).
- level-1: The agent determines the level of the detail in which the data is presented (e.g., one user is given a cumulated overview whereas another one is presented a detailed line-up of concrete data).
- level-2: The agent states the rights and tools the specific user is able to use on the data presented (e.g., some members of the team may have the right to modify a design, others is only given the possibility to view and comment it).
- level-3: The agent is used to provide an individual help function based on hypermedia links (e.g., the documents generated include links to additional help files that are chosen in respect to the knowledge of the specific user).

All these levels can be implemented individually or might be combined to achieve higher flexibility and effect.

5.2   User Model and a Concept of Intelligent Agents

The adaptive component of the information system uses information stored in a special group/user model. As shown in the architectural model above, the intelligent agent accesses this information and generates the suitable representation for the data to be retrieved.

The group/user model is a hierarchically organized data collection, that provides a bunch of information units to each user. For instance, preferred representation form for different kinds of data, initial level-of-detail for different kinds of data, expert levels for different presentation tools, preferred screen layouts, and hardware capabilities of the users workstation (e.g., to determine if some presentation forms like audio can be used automatically or have to be confirmed individually), etc.

The hierarchical organization of these data allows the definition of clusters (groups) of individual users, for which a default user group profile is generated. Each user may change some or all information of the group profile in his personal user profile. Then, the intelligent agent performs the following steps to determine the representation suitable for a specific user:

```
Check user profile,
User profile exist,
    agent finds specific attribute/value combination,
    IF attribute/value combination is found,
    THEN end scanning,
         start processing,
    ELSE scan group profile,
         agent finds specific attribute/value combination,
         IF attribute/value combination is found,
         THEN end scanning,
              start processing,
         ELSE neither such combination is found,
              use overall default value
```

Changes in the user profile can be divided to modifications that the user is allowed to perform by himself and others that have to be verified by a system administrator, i.e., individual user is never able to change access rights himself.

The aforementioned approach developed for the ASN has been evaluated by an application that supports Rapid Prototyping (RP) by providing an automatic selection of production procedures [9]. The graphical user interface (GUI) of this system is shown in Fig. 4.

The GUI allows to specify the design requirements by some parameters of the product (e.g., its hardness and compactness). After specification of these parameters, the tool selects the best RP procedure to generate a prototype of the product. Afterwards, the input and output parameters of this selection are stored in the knowledge base in order to reuse them later.

When the product requirements change during the design process and new parameters are stored in the knowledge base, the ASN will automatically start two

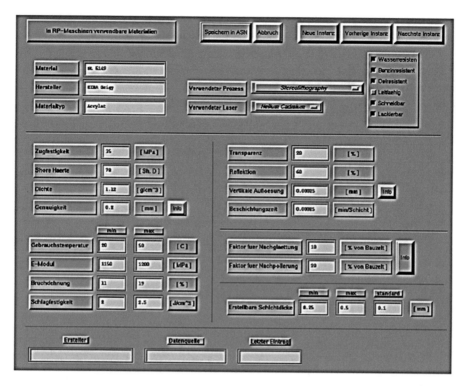

Fig. 4. Graphical user interface of Rapid Prototyping selector [9]

different rules. First, the ASN re-start the selection of RP procedures and the new results of this selection are re-stored in the system database. Second, the corresponding designers are informed by means of communication about the changed results (e.g., electronic mail) and asked to react upon the new situation and checking if their prototype still fulfil the requirements.

5.3   Generating Rapid Prototyping Procedure Using Genetic Algorithm

To generate rapid prototyping (RP) procedure, the genetic algorithm structure can be used, it is by dividing the surface of a modelled part into a grid and treating each cell of the grid as a node in the travelling salesman problem. By comparing the starting surface to the end product surface at each depth, an RP procedure can be generated.

To design the genetic algorithm, the nature of the modelled part has to be converted into a genetic structure. To represent an RP procedure in a genetic form, each coordinate point's location are stored as the genotype of a gene in a chromosome. A chromosome in this context represents a particular sintering route through all the available coordinate points. Then, each species in a population will represent a possible solution to the problem of finding the most efficient sintering route.

14   D. Roller and B. K. Ane

As illustration, a case study is adopted from [10]. Figures 5 and 6 illustrate an example part and how the grid system can be applied to the modelled part. Figure 7 illustrates the structure of the ordered chromosome. Each number in the genotype refers to one of the nodes (i.e., coordinate points) in the matrix shown in Fig. 6. To create a sintering route from the data in the chromosome, the nodes have to be joined in a linear fashion in the order described by the chromosome as depicted in Fig. 8.

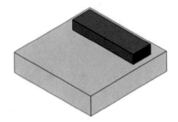

20	21	22	23	25
15	16	17	18	19
10	11	12	13	14
5	6	7	8	9
0	1	2	3	4

Fig. 5. 3D model of part [10]

Fig. 6. The grid of modelled part.

Gene	0	1	2	3	4	5	6	7	8	9	10	11	12	13	14	15	16	17	18	19	20	21	22	23	24	25
Genotype	0	1	2	3	4	5	11	17	23	24	18	12	6	7	8	9	10	16	22	25	19	13	14	15	21	20

Fig. 7. Example chromosome for grid of the modelled part.

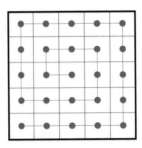

Fig. 8. Example sintering route for chromosome in Fig. 7.

Fitness Function
To assess the quality of each species, a fitness function is used. The fitness function compares one or more features in the species to one or more specific goals. Here the goal is to generate a sintering route with the shortest length possible while being as straight as possible.

The length of the sintering route which sums the distance between all of the ordered nodes in the species' route is calculated using the following equation:

$$\text{Total route length} = \sum_{n=1}^{p} \sqrt{(x_n - x_{n-1}) + (y_n - y_{n-1})} \tag{1}$$

where p: number of nodes, x: position in x axis, and y: position i y axis.

In order to keep the route as straight as possible, a second equation is inserted into the fitness function to reward continuation of a previous direction:

$$\text{Countinuous route length} = \frac{1}{2^{p-1}} \tag{2}$$

where p: number of consecutive points along same direction.

Gene Crossover

To create a new generation of species, a selection of the fittest species from the previous generation need to share their genes with each other, so that allows the population to create new chromosomes by retaining superior gene combinations (the elitists) while discarding inferior combinations. In this context, crossover is considered as the most important search process. During crossover, firstly, a pair of parents is randomly selected from the mating pool. Secondly, a point, called crossover site, along their common length is randomly selected, and the information after the crossover site of the two parent strings are swapped, thus creating two new offsprings.

Here, the sequential constructive crossover (SCC) operator is applied that constructs an offspring using better edges on the basis of their values present in the parents' structure. The SCC also uses the better edges, which are present neither in the parents' structure. It does not depend only on the parents' structure as depicted in Fig. 9. The SCC sometimes introduces new, but good, edges to the offspring which are not even present in the present population. Hence, the chances of producing a better offspring are more than the other methods such as the edge recombination crossover (ERC) and generalized N-point crossover (GNC) [11].

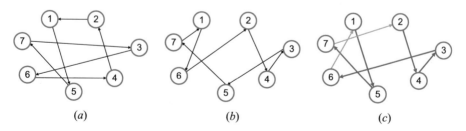

(a)   (b)   (c)

Fig. 9. Example of sequential constructive crossover operator: (a) P1: [1, 5, 7, 3, 6, 4, 2], (b) P2: [1, 6, 2, 4, 3, 5, 7], and (c) Offspring: [1, 5, 7, 2, 4, 3, 6] [11]

Mutation

To ensure the genetic algorithm converges on an optimal solution, it is important to keep the available genes in the gene pool as diverse as possible. One way to do this is to insert new gene combinations into the gene pool by adding a mutation function into the genetic algorithm. As the algorithm is dealing with an ordered chromosome, mutating a gene value to a random number can cause a node to be visited twice and another node not at all. To avoid this, a method of mutating the current gene to one of the nearest neighbours of the previous gene is used. This method calculates the nearest neighbours of the previous gene and randomly selects one of the available neighbours.

Figure 10 is an example of a sintering route which is not completely optimal. In Fig. 11, the gene targeted for mutation is highlighted along with its nearest neighbours. The genetic algorithm will then select one of these nearest neighbours at random to become the next node in the route order. In Fig. 12, the optimal nearest neighbour is selected and its position in then switched with the original gene to ensure order is kept in the chromosome. Figure 13 exhibits the result of the mutation where the two switched genes are highlighted and the sintering route is now of optimal length and direction.

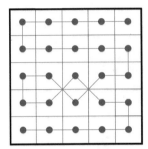   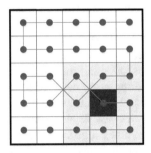

Fig. 10. Non optimal route   **Fig. 11.** First gene location with nearest neighbors

6 Illustrative Example: Constraint Propagation in Mechanical CAD

In order to provide an illustration, an application taken from the object-oriented programming environment MOSS (Multi-level Object Simulation System) in the field of mechanical CAD [12] is provided.

MOSS is a result from the implementation of property driven model (PDM). The PDM representation is based on the concept of property (attributes and relationships). PDM uses a four-layer specification hierarchy, i.e., instance layer, model layer, meta-model layer, and meta-meta-model layer. It is a consistent object system, whereas every object has a unique identifier and there are no dangling references. The overall MOSS architecture is shown in Fig. 14.

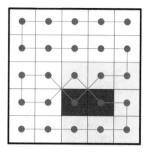
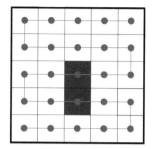

Fig. 12. Nearest neighbor selected

Fig. 13. Next gene location switched with nearest neighbor

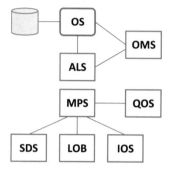

OMS: Object Manager

ALS: Augmented LISP System

MPS: Message Passing System

QOS: Query Optimizer

SDS: Static Definition System

LOB: Multi-user Interactive Object editor

IOS: I/O Sub-System

Fig. 14. Overall architecture of MOSS system [12]

Versioning

The versioning mechanism is very useful in the domain of CAD. Versioning can be used for exploring parallel hypotheses, or for doing concurrent design, letting design teams diverge, then reconciliating the separate designs [13].

In the MOSS system all objects are versioned, including classes and methods. However, although consistency is enforced between classes and instances, there is no provision for a mechanism associating new classes with old instances, or new instances with old versions of classes.

In the MOSS system, the end user has no control over the version mechanism. A new version is automatically created at each new general update. A new revision can be created by collapsing all information onto a single node. The other nodes can then be archived.

Concurrence

Concurrence is used to allow several users to work simultaneously on the same information. MOSS uses an optimistic concurrence approach, i.e., when an update is performed; all operations are checked for consistency with the previous operations. If none of the integrity constraints is violated, then the version is added linearly to the version graph. If some inconsistency is found, then the user is notified and may elect either to modify the data to comply with the recorded modifications, or to create an alternative database version.

6.1   Representation of Mechanical Objects

Because of the important features of MOSS such as semantic representation, versioning, persistency, concurrence, MOSS is quite suitable for the representation of mechanical objects [14], the representation of design knowledge [15], and even the representation of messages and agents in the distributed integrated design environment [16].

Here MOSS is used the design environment for representing mechanical parts and for propagating constraints. Considering a simple mechanical assembly consisting of two plates fixed by means of a bolt, a nut and a washer (see Fig. 15). This assembly can evidently be represented as in Fig. 16 using hierarchical relations, and can also be represented completely as in Fig. 17 using semantic relations.

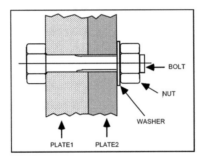

Fig. 15. Mechanical assembly [12]

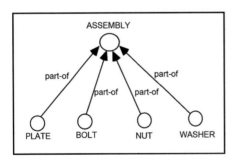

Fig. 16. The hierarchical relations among the classes [12]

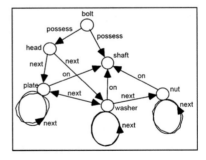

Fig. 17. The semantic relations among the classes [12]

The definition interface of MOSS allows us to create classes simply. For example, the class WASHER can be defined as follows:

```
(defclass WASHER SWASH test
(:sp PART-OF ASSEMBLY)
( : tp THICKNESS)
(:tp INTERNAL-DIAMETER)
(:tp EXTERNAL-DIAMETER)
(:sp WASHER-TYPE WASHER-TYPE)
(:sp NEXT NUT WASHER»
```

A washer is defined with three terminal properties: thickness, internal diameter (standard) and external diameter (standard); it is related to other classes through the structural properties: washer type (correspondent to a standard table), composition relation PART-OF and the possible relation NEXT with the matching parts.

6.2   Constraint Propagation in Mechanical CAD Systems

Here, a constraint propagation technique, namely, along a local semantic link such as NEXT to the matching parts has been applied due to its capability that allows the components of the system to be presented clearly by using semantic links and constraint verification can be made while the objects are being modified.

If a mechanical assembly has to be modified, e.g., the bolt diameter too thick or too thin, then it can be done simply by sending a message:

```
(send <bolt> '=modify-diameter <new-value»
```

In the system, the modification propagates automatically to the "next" part by a new message contained in the method = *modify-diameter* for the class BOLT:

```
(send (g==> 'HAS-NEXT '=modify-value <~ >1).
```

In this process, propagation continues as new messages are sent to other objects.

7   Conclusion

Mechatronic systems comprise of mechanical and electrical components. Mechanical design changes lead to design modifications on the electrical side and vice versa. The contemporary computer-aided environments are unable to provide a sufficient degree of integration in order to allow the bi-directional information flow between the MCAD-ECAD domains. Thus, overcoming this barrier of systems integration will release a tremendous efficiency potential with regard to the development of mechatronic product platforms.

A key challenge in the creation of an integrated solution for interdisciplinary product design that involves the usage of MCAD and ECAD systems is the need to capture all the data used in the process. This is a difficult challenge since data can be lost or miss-interpreted if one of the systems in the process cannot properly represent or manage data received from the other system.

This paper proposes an approach to achieve integration of mechanical and electrical CAD systems through the modelling and propagation of constraints. Cross-disciplinary constraints between mechanical and electrical design domains are classified, represented, modelled, and bi-directionally propagated in the active semantic network (ASN) in order to provide immediate feedback to designers of both engineering domains. The ASN has been identified as a potential solution that supports the integration of MCAD and ECAD systems. It works by seeking commonality amongst the solutions in order to provide a common representation of data that can be shred by the MCAD and ECAD systems.

However, it is noteworthy that the solutions currently available mainly focus on solving specific problems faced by particular CAD vendors. The direction for future research with regard to the integration of mechanical and electrical CAD will be anchored in answering such open questions as: (*i*) Robustness – *how can such a metric be changed from one application to another without significant loss of accuracy or semantics?*, (*ii*) Error handling – *can the system automatically correct and document errors?*, and (*iii*) Human cognition and human factors – *to what degree can the systems help engineers interpret data and suggest solutions?* It is important to identify connections between these solutions and an understanding as to how each solution fits into the whole picture of MCAD–ECAD Integration.

References

1. Rudeck, E.: An Introduction to Parametric Modelling, Concurrent Engineering (2013)
2. Chang, K.-H.: e-Design: Computer-Aided Engineering Design. Academic Press (2015)
3. Chen, K., Bankston, J., Panchal, J.H., Schaefer, D.: An integrated approach to design mechatronic systems: cross-disciplinary constraint modelling. In: Proceedings of the ASME 2008 International Design Engineering Technical Conferences & Computers and Information in Engineering Conference (IDETC/CIE), New York, USA (2008)
4. Chen, K., Bankston, J., Panchal, J.H., Schaefer, D.: A framework for integrated design of mechatronic systems. In: Wang, L., Nee, A.Y.C. (eds.) Collaborative Design and Planning for Digital Manufacturing. Springer, London (2009)
5. Schaefer, D., Eck, O., Roller, D.: A shared knowledge base for interdisciplinary parametric product data models in CAD. In: Proceedings of the 12th International Conference on Engineering Design, pp. 1593–1598 (1999)
6. Roller, D., Eck, O., Dalakakis, S.: Advanced database approach for cooperative product design. J. Eng. Des. **13**(1), 49–61 (2002)
7. Behbahani, S., De Silva, C.W.: Mechatronic modeling and design. In: De Silva, C.W. (ed.) Mechatronic Systems: Devices, Design, Control, Operation and Monitoring, pp. 17-3–17-10. CRC Press (2008)
8. Roller, D., Eck, O.: Active cooperative transaction model for shared design databases. In: Proceedings TeamCAD: GVU/NIST Workshop on Collaborative Design, Atlanta, pp. 193–197 (1997)
9. Roller, D., Bihler, M., Eck, O.: ASN: active, distributed knowledge base for rapid prototyping. In: Proceedings of 30th ISATA, Volume "Rapid Prototyping in the Automotive Industries and Laser Applications in the Automotive Industries". Automotive Automation Ltd., Croydon, pp. 253–262 (1997)

10. Essink, W.P., Nassehi, A., Newman, S.T.: Toolpath generation for CNC milled parts using genetic algorithms. In: Zaeh, M.F. (ed.) Enabling Manufacturing Competitiveness and Economic Sustainability. Springer (2014). ISBN: 978-3-319-02053-2
11. Ahmed, Z.H.: Genetic algorithm for the traveling salesman problem using sequential constructive crossover operator. Int. J. Biometrics Bioinf. (IJBB) **3**(6), 96–105 (2010)
12. Shen, W., Barthes, J.-P.: Description and applications of an object-oriented model PDM. In: Dubois, J.-E., Gershon, N. (eds.) Modelling Complex Data for Creating Information, pp. 15–24. Springer, Heidelberg (1996)
13. Barthès, J.P.: La problèmatique de reconciliation en ingénierie simultanée, Actes de 01 DESIGN 1993 (1993)
14. Finger, S., Rinderle, J.R.: Representation of mechanical design. In: Proceeding of The Third Workshop on Intelligent CAD, IFIP Working Group 5.2, Osaka, Japan (1989)
15. Shen, W., Barthès, J.P.: Toward a multi-agent architecture for distributed integrated design environment (DIDE), Technical report 94-1, CNRS URA 817 Heudiasyc, UTC (1994)

Experimental Fuzzy Models

Experimental FuzzyWA Aggregated Location Selection Model for Very Large Photovoltaic Power Plants in Global Grid in the Very Early Engineering Design Process Stage

Burak Omer Saracoglu$^{(\boxtimes)}$

Orhantepe Mahallesi, Tekel Caddesi Istanbul, Turkey
burakomersaracoglu@hotmail.com

Abstract. Our World has been polluted, mostly after the industrialization, because of not taking into account the environment and other species. Now, our World gives vital signs. Some actions should be taken to stop the activities that affect the climate catastrophically. One of the important activities in this respect is to model, build and operate a 100% renewable electricity grid on the World (Globalgrid). This target isn't so easy to be achieved without a very well organized and planned system. One of the major renewable energy source in the Globalgrid will be the solar energy. The very large photovoltaic power plants will most probably play a critical role. This paper presents an experimental fuzzy weighted average aggregated location selection model for the very large photovoltaic power plants in the Globalgrid Concept in the very early engineering design process stages to help the modeling, designing, constructing and operating efforts of the Globalgrid.

Keywords: Fuzzy logic · Climate change · Concept · Design
Earth · Electricity · Energy · Engineering · Engineering design process
FuzzME · Fuzzy Models of Multiple-Criteria Evaluation · Fuzzy expert system
Fuzzy weighted average · Globalgrid · Globalgrid concept
Multiple-criteria fuzzy evaluation · Photovoltaics · Photovoltaic power plants
Solar power · Very large photovoltaic power plants

1 Introduction

Our World, *The Earth: the only planet that the humankind can live today*, has been repeatedly giving some very somber, significant and vital signs and warnings for more than a decade, that humankind has been killing, poisoning, polluting and contaminating the environment and other species (e.g. Diademed Sifaka, and Pere David's Deer visit [1, 2]) for years and years; mostly and largely after the industrialization, because of not taking into account the environment and other species, not obeying the rules of nature, not having any respect to the other human beings, and to itself (Homo sapiens), the environment and other species and insisting on only the maximization of profit (only concerning money, more money, more and more money), but nothing else. Some of these vital signs and clues, such as increase in global mean temperatures, melt in the

© Springer Nature Switzerland AG 2019
B. K. Ane et al. (Eds.): WSC 2014, AISC 864, pp. 25–35, 2019.
https://doi.org/10.1007/978-3-030-00612-9_3

Arctic Sea Ice, decline in the Greenland ice sheets, ocean warming, sea level rising, increase in extreme weather events (droughts, floods, snowfalls, heat waves, etc.), have been observed, recorded and published by several international organizations and very fair-minded and non-partisan researchers [3–7], which shows us that future will be very difficult for next generations of humankind. Although, nobody can be sure nowadays and none of the research studies can be specific on the actions, which have been taken and will be taken, are already very lately or not; taking actions immediately will always be better than taking no actions in this issue (Has the humankind been already too late or not?). There are several improvements and activities to be done in several sectors to stop the catastrophic climate effects' causes, for instance in the electricity generation sector %100 renewable electricity (RE) grid, in the transportation sector the public transportation or the electric vehicles (EV). In this respect, the solar power and the very large photovoltaic power plants (VLPVPPs) in the Globalgrid concept will most probably play a key role in the electricity generation sector. The definition of VLPVPPs are not clear and sharp yet, however in this study they are accepted as the photovoltaic solar power plants that have the installed power of 1.000 MW_p (peak power) or more. Two photovoltaics power plants in operation can be seen in Fig. 1 to better understand this technology and these kings of power plants.

265 MW in Calexico, California, USA   6 MW, AJO, Arizona, USA

Fig. 1. Two photovoltaics power plants in operation (Source: [8] p. 42 (left) & Array Technologies documents (right) on http://arraytechinc.com/ credit for both: Array Technologies).

The possibility, designability, feasibility, effectiveness, efficiency, usability, and operability of the Globalgrid shall only be possible by finding, designing and investing on the appropriate VLPVPPs' locations (also for other power plant types). Otherwise, the system can't be operational. In the current study, an experimental fuzzy weighted average (fuzzy WA) aggregated location selection model for the VLPVPPs in the Globalgrid Concept in the very early engineering design process stages is presented. The next section presents the previous studies in the literature. The experimental proposed fuzzy weighted average (fuzzy WA) aggregated model and its application is explained in the Sect. 3. The concluding remarks and further research studies are presented in Sect. 4.

2   Literature Review

A detailed literature review was tried to be conducted on the academic publication online database and journals (ACM Digital Library [9], Directory of Open Access Journals [10], Google Scholar [11], Science Direct [12], Springer [13]) for the key terms and phrases of *"very large photovoltaic solar power plant" and "fuzzy weighted average", "very large photovoltaic solar power plant" and "fuzzy WA", "very large photovoltaic solar power plant" and "FuzzME", "very large photovoltaic solar power plant" and "Fuzzy Models of Multiple-Criteria Evaluation", "photovoltaics" and "fuzzy weighted average", "photovoltaics" and "fuzzy WA", "photovoltaics" and "FuzzME", "photovoltaics" and "Fuzzy Models of Multiple-Criteria Evaluation", "Globalgrid" and "fuzzy weighted average", "Globalgrid" and "fuzzy WA", "Globalgrid" and "FuzzME", and finally "Globalgrid" and "Fuzzy Models of Multiple-Criteria Evaluation"* as presented in Table 1.

Table 1. Literature review summary of the current study (final date of the review on [9–13]: 06th October 2014; "renewable" added during the revision).

Key terms and phrases	No. of documents
"very large photovoltaic solar power plant" and "fuzzy weighted average"	0
"very large photovoltaic solar power plant" and "fuzzy WA"	0
"very large photovoltaic solar power plant" and "FuzzME"	0
"very large photovoltaic solar power plant" and "Fuzzy Models of Multiple-Criteria Evaluation"	0
"photovoltaics" and "fuzzy weighted average"	0
"photovoltaics" and "fuzzy WA"	0
"photovoltaics" and "FuzzME"	0
"photovoltaics" and "Fuzzy Models of Multiple-Criteria Evaluation"	0
"Globalgrid" and "fuzzy weighted average"	0
"Globalgrid" and "fuzzy WA"	0
"Globalgrid" and "FuzzME"	0
"Globalgrid" and "Fuzzy Models of Multiple-Criteria Evaluation"	0
"renewable" and "fuzzy weighted average"	0
"renewable" and "fuzzy WA"	0

This literature review concluded that the fuzzy weighted average had not most probably been applied in any very large photovoltaic solar power plants studies until 06th October 2014. In addition, the FuzzME software had not been used for any studies on the very large photovoltaic solar power plants until 06th October 2014. Based on these findings, it could be underlined that this study would be one of the first studies applying the fuzzy weighted average method in the location selection model for the VLPVPPs in the Globalgrid Concept in the very early engineering design process stages. Henceforth, this study had some unique and very different challenges such as finding, defining, identifying and selecting the factors for the location selection model of the VLPVPPs in the Globalgrid Concept in the very early engineering design process stages and considering the different perspectives of weight assignments in the model. Although the International Energy Agency Photovoltaic Power System Programme had finalized a task (Task 8) for the very large scale photovoltaic power generation (VLS-PV) systems (see [14]), the scope and the approach of this study (such as the Globalgrid Concept, the VLPVPPs) and the previous studies could not be categorized as same and similar. They focused on the design possibility, the requirements analysis and the applicability in the deserts. On the other hand, the author gathered data and information from these studies and built and developed most of the pillars of this study based on those very important research studies on photovoltaics. This literature review presents that this study is most probably the first study in its broader subject (renewable energy).

3   The Experimental Proposed Fuzzy Weighted Average (Fuzzy WA) Aggregated Model

The experimental proposed fuzzy WA aggregated model in this study is based on the special and distinctive main factors and sub factors, that are very convenient for the location selection model of the VLPVPPs in the Globalgrid Concept in the very early engineering design process stages, as the Global Horizontal Irradiance (GHI) (Criteria 1: C_1), the HVDC & HVAC electricity grid infrastructure (C_2) (HVDC: High-Voltage Direct Current, HVAC: High-Voltage Alternating Current), that consists of two sub-factors as the HVDC transmission system network (C_{21}) and the HVAC transmission system network (C_{22}), the security conditions (C_3), that consists of four sub-factors as the terrorism status (C_{31}), the sabotage status (C_{32}), the free travel status (C_{33}) and the social chaos status (C_{34}), the war conditions (C_4), that consists of two sub-factors as the war status (C_{41}) and the war predictions for near/short to mid terms (C_{42}), and finally the transportation (C_5), that consists of five sub-factors as the air transportation (C_{51}), the railway transportation (C_{52}), the road transportation (C_{53}), the inland water way transportation (C_{54}), and the ocean-sea transportation (C_{55}).

One of the important and necessary factors in this model is the GHI, because of being one of the main characteristics elements of this electricity generation technology such that: The energy generated by a PV system is calculated by the general equations

$$E_{grid} = E_A \times \eta_{inv} \tag{1}$$

where E_{grid} is the energy available to the grid, E_A is the array energy available to the load, η_{inv} is the inverter efficiency,

$$E_A = E_P \times (1 - \lambda_p) \times (1 - \lambda_c) \tag{2}$$

where E_P is the energy delivered by the PV array, λ_p is the miscellaneous PV array losses, λ_c is the other power conditioning losses,

$$E_P = S \times \eta_P \times \bar{H}_t \tag{3}$$

where S is the area of the array, η_P is the average efficiency of the array, $\bar{H}_t$ is the daily total tilted irradiance related with the global horizontal irradiance (look at [15] pp. PV.23–PV.21) (see [16] pp. 20–22 and [17] for these kinds of equations). Hence, the GHI criteria (C_1) is one of the indispensable factors for the location selection model of the VLPVPPs in the Globalgrid Concept. The GHI is defined and explained clearly by the National Renewable Energy Laboratory (NREL) [18]. Accordingly, the location selection model of the VLPVPPs has to include this factor as one of the primary factors.

This model was directly built on the FuzzME Software. This software was developed to handle various methods for the aggregation of partial evaluations such as the fuzzy weighted average, the Fuzzy OWA (weighted average/weight averaging) operator, and the Fuzzified WOWA (fixed weights and order weights) operator by Holecek, Talasova, Pavlacka and Bebcakova (see [19, 20] for FuzzME). The importance and difficulties of the designing and modeling of the linguistic terms, scales, and hedges of the factors were very well known and experienced by the author, so that the previous research studies were investigated in a detailed manner. In this respect, some of the researchers and academics, who contributed very much, such as Rensis Likert (1903–1981) (5 Likert scales), Lotfali Askar Zadeh (1921–alive) (fuzzy set, fuzzy logic, and hedges), George Armitage Miller (1920–2012) (magical number 7), Richard M. Shiffrin (1968–alive) and Robert M. Nosofsky (alive) (magical number 7, 7 ± 2 rule) and finally Bernd Rohrmann (alive), had to be aware of and their recommendations had to be bear in mind during these kinds of studies (see [21–25]. By this attitude, approach and modeling philosophy, the reality of the fuzziness (terminology by Zadeh for the unsharpness of class boundaries) of the human concepts and human reasoning could be very much easily understood, accepted, and applied to solve the real world problems. The main aim in the modeling philosophy of the solution of this problem is developing mostly agreed upon, simple, understandable, applicable and the least costly model. During the definitions and the decisions on the linguistic terms and scales of the factors C_3, and C_4, the risk of attack scale of the Homeland Security Advisory System by the U.S. Department of Homeland Security was taken into consideration and adopted to the current model (only for linguistic terms) (see [26, 27]).

The aggregation of the partial evaluations in this study were directly done by the fuzzy weighted average operator on the FuzzyME Software. The importance of the factors could not be decided with ease in the sharp ways, so that the fuzzy numbers could give a hand to the modeling of the real world application by the fuzzy weighted average. The fuzzy WA, its computations and applications were presented by several researchers and academics (see [28–33]), which helped to organize this one in a good manner (see Table 2). During the weight assignment of the nodes on the goal or decision tree, the uniform weights were first created for all of the sub-nodes by the help of the FuzzME, then the adjustments were directly made to finalize the fuzzy weights of the nodes on the FuzzME. All of the warning messages such as "The weights are not correct" and, "The weights are not normalizable" had been followed one by one until none of the warnings was appeared. Whenever the normalization warning appeared, the normalization tab was selected and the software recommendation were evaluated on the

FuzzME (see Table 3). After all of the fuzzy weights were assigned the evaluations were started to be performed for the randomly selected alternatives on the world.

Table 2. The representative normalized membership functions on the FuzzME (open the presentation file and the FuzzME model file).

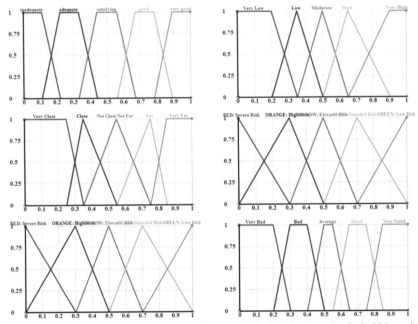

* Note 1: Goal: Increasing Scale: higher values are better (up left), GHI C_1: Increasing Scale: higher values are better (up right), HVDC & HVAC C_2: Decreasing Scale: lower values are better (middle left), Security Conditions C_3: Increasing Scale: higher values are better (middle right), War Conditions C_4: Increasing Scale: higher values are better (down left), Transportation C_5: Increasing Scale: higher values are better (down right).
* Note 2: Increasing Scale: values in the real life are normalized with the maximum value, Decreasing Scale: first, values in the real life are normalized with the maximum value; second, normalized values are calculated and changed as x to (1-x) or directly decreasing scale selected on the FuzzME.

An experimental application were performed in the current study. Different provinces, districts and regions of different countries were tried to be analyzed and evaluated by help of the current model on the FuzzME Software. The location alternatives were considered and assumed to be in Spain, in Greece, and in Turkey in Europe (3 alternatives), in the USA, in Mexico, in Peru, and in Chile in America (4 alternatives); in Mauritania, in Mali, in Niger, in Chad, and in Sudan in Africa (5 alternatives); in Afghanistan, in Pakistan, in India and in China in Asia (4 alternatives); in Papua New Guinea, in Tasmania, in Australia (3 alternatives) without taking into account some important main features and issues such as the continental differences and the electricity demand (totally 19 alternatives). All of the alternatives were evaluated on simple scale

Table 3. The weights of fuzzy WA on the FuzzME (open the presentation file and the FuzzME model file).

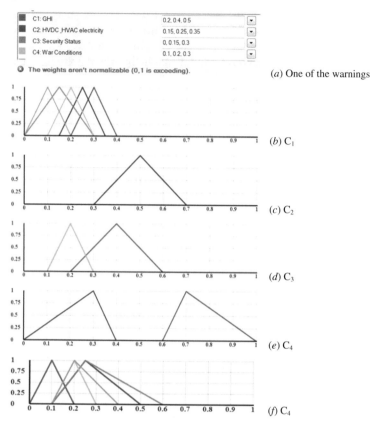

(a) One of the warnings

(b) C₁

(c) C₂

(d) C₃

(e) C₄

(f) C₄

type (in the node/scale type menu/FuzzME) for each the criterion value in this experimental application study. The alternatives were evaluated on the basis of the continental groups (not all at once) for diminishing and decreasing the mental workload and taking into consideration the cognitive issues (3 + 4 + 5 + 4 + 3 = 19 alternatives) (see [23, 24]). The data and information for these alternatives on each criterion were gathered from several official websites and documents and also from unofficial websites and documents. For instance, the boarders of the countries were learnt from [34], the GHI was evaluated based on [35], the main Globalgrid transmission system network approximation were taken from [36], the road transportation evaluation was performed by help of [37], and the ports data and information were taken from [38]. The evaluations were made by only one decision maker with very limited data and information available on the decision maker's hands.

While the evaluations for each alternative were made by the decision maker, the computations for all of the alternatives were directly performed by the FuzzME, and after all of the evaluations of all of the alternatives had been finalized, the results were directly taken as the sorted alternatives on the FuzzME (sort by evaluation tab). In the current experimental study, the sort of the alternatives were the Alternative 19, the Alternative 5, the Alternative 6, the Alternative 7, the Alternative 16, the Alternative 17, the Alternative 2, the Alternative 4, the Alternative 18, the Alternative 8, the Alternative 9, the Alternative 1, the Alternative 15, the Alternative 12, the Alternative 10, the Alternative 11, the Alternative 14, the Alternative 13, and the Alternative 3 (see Table 4). The Alternative 19 was in the good to very good set. The good set had the alternatives of 5, 6, 7, 16 and 17.

Table 4. The representative normalized membership functions on the FuzzME (open the presentation file and the FuzzME model file).

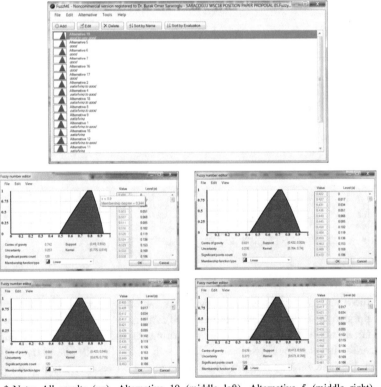

   * Note: All results (up), Alternative 19 (middle left), Alternative 5 (middle right), Alternative 6 (down left), Alternative 7 (down right)

   The satisfying to good set had the alternatives of 2, 4, 18, 8, 9, and 1. The satisfying set had the alternatives of 15, 12, 10, 11, 14, and 13. The Alternative 3 was in the adequate to satisfying set. Based on this finding, in a study in the real world application, the concentration and the efforts of the researchers, the engineers, the analysts, the investors, the governments and the international organizations and bodies had to be directed into the good to very good set (Alternative 19) and then into the good set (Alternatives of 2, 4, 18, 8, 9, and 1) and finally into the satisfying set (Alternatives of 15, 12, 10, 11, 14, and 13) for selecting the most satisfying and advisable location of the very large photovoltaic power plants in the Globalgrid Concept in the very early engineering design process stages. These experimental results indicate that the good to very good set contains the most appropriate locations. The further detailed studies should first be started for this set. Afterwards, the good set should be taken into account and the further detailed studies should be performed for this set. These experimental results mainly mean that if and when the detailed analysis of the locations are started in the order of the ranked locations as performed in the way of this experimental study, the most satisfactory locations will be found and analyzed in the Globalgrid by the least spent resources (time. money, etc.). The best Globalgrid design will be achieved by considering the studies developed based on these kinds of the experimental studies and their findings and results.

4   Conclusions and Future Work

It is believed that this study can present to the readers, the researchers, the investors and the international organizations and bodies, the easiness, the usability and the adopt-ability of the weighted average (WA) operator with the fuzzy set and logic combination to build the fuzzy WA applications for the solution of the location selection problem for the very large photovoltaic power plants (VLPVPPs) in the Globalgrid in the very early engineering design process stages by help of the FuzzME. It is hoped that the more sophisticated models based on this one shall be developed and help the electricity generation sector researchers, engineers and investors (in this case PV) to start to investigate the possibility of the engineering, the construction and the operation of the VLPVPPs in the Globalgrid. The current model in this study should be improved in several issues such as the location selection factors of the VLPVPPs and the utility-scale PV power stations has to be identified and defined in a detailed manner, the fuzzification of the identified and defined factors has to be specifically studied in detail, moreover, the weights of the fuzzy WA operators should carefully be studied and finally the group decision making model should be developed for conducting the world wide case studies to help the international organizations, the multinational foundations, the governments and the investors.

Acknowledgments. The author would like to thank to Dr. Bernadetta Kwintiana Ane (conference), and Dr. Pavel Holeček (FuzzME). This study shall never be finalized and submitted to the conference without their consideration, guidance, and help. Please send your comments, feedbacks and criticisms to my e-mail (burakomersaracoglu@hotmail.com) in any format at any time. Your feedback will be very important and valuable for me during the development process of the models and systems for the real life applications.

References

1. The International Union for Conservation of Nature and Natural Resources, The IUCN Red List of Threatened Species, Diademed Sifaka (English common name), Propithecus diadema (scientific name). http://www.iucnredlist.org/details/summary/18358/0
2. The International Union for Conservation of Nature and Natural Resources, The IUCN Red List of Threatened Species, Pere David's Deer (English common name), Elaphurus davidianus (scientific name). http://www.iucnredlist.org/details/7121/0
3. Young People's Trust for the Environment. http://www.ypte.org.uk/environmental/our-polluted-planet/34
4. World Wildlife Fund. http://www.worldwildlife.org/threats/pollution
5. Union of Concerned Scientists. http://www.ucsusa.org/global_warming/science_and_impacts/science/global-warming-faq.html#.VCkSvU0cQkI
6. Intergovernmental Panel on Climate Change (IPCC). http://www.ipcc.ch/publications_and_data/publications_and_data.shtml
7. Emanuel, K.: Increasing destructiveness of tropical cyclones over the past 30 years. Nature **436**(7051), 686–688 (2005)
8. Renewables Insight-Energy Industry Guides (RENI), PV Power Plants 2014 Industry Guide, European edition (2014) & Array Technologies documents on http://arraytechinc.com/
9. ACM Digital Library. http://dl.acm.org
10. Directory of Open Access Journals. http://doaj.org
11. Google Scholar. http://scholar.google.com.tr
12. Science Direct. http://www.sciencedirect.com
13. Springer. http://www.springer.com/?SGWID=5-102-0-0-0
14. The International Energy Agency Photovoltaic Power System Programme, Very Large Scale PV Power Generation Systems. www.iea-pvps.org
15. RETScreen® International, Clean Energy Project Analysis: RETScreen® Engineering & Cases Textbook, Photovoltaic Project Analysis Chapter, Clean Energy Decision Support Centre, Minister of Natural Resources Canada (2001)
16. Carl, C.: Calculating solar photovoltaic potential on residential rooftops in Kailua Kona, Hawaii, Master of Science Geographic Information Science And Technology, Faculty of the USC Graduate School University of Southern California (2014)
17. Gilman, P.: System Advisor Model SAM Photovoltaic Model (pvsamv1) Technical Reference, Draft (2014)
18. Global Horizontal Radiation, Global Horizontal Irradiance, Glossary of Solar Radiation Resource Terms, The National Renewable Energy Laboratory (NREL), The U.S. Department of Energy. http://rredc.nrel.gov/solar/glossary/gloss_g.html
19. Talasova, J., Holecek, P.: Multiple-criteria fuzzy evaluation: the FuzzME software package. IFSA/EUSFLAT Conf. **2009**, 681–686 (2009)
20. The website of FuzzME. http://www.fuzzme.net, http://fuzzme.wz.cz
21. Likert, R.: A technique for the measurement of attitudes. Archives of Psychology, No: 140, New York, USA (1932)
22. Zadeh, L.A.: A fuzzy-set-theoretic interpretation of linguistic hedges. J. Cybern. **2**(3), 4–34 (1972)
23. Miller, G.A.: The magical number seven, plus or minus two: some limits on our capacity for processing information. Psychol. Rev. **63**, 81–97 (1956)
24. Shiffrin, R.M., Nosofsky, R.M.: Seven plus or minus two: a commentary on capacity limtations. Psychol. Rev. **101**(2), 357–361 (1994)

25. Rohrmann, B.: Verbal qualifiers for rating scales: Sociolinguistic considerations and pschometric data, Project Report, University of Melbourne/Australia (2007). http://www. rohrmannresearch.net/pdfs/rohrmann-vqs-report.pdf
26. The Department of Homeland Security, Citizen Guidance on the Homeland Security Advisory System. www.dhs.gov/xlibrary/assets/citizen-guidance-hsas2.pdf
27. The Department of Homeland Security, Chronology of Changes to the Homeland Security Advisory System. http://www.dhs.gov/homeland-security-advisory-system
28. Broek, P.V.D., Noppen, J.: Fuzzy weighted average - analytical solution. In: Proceedings of the International Joint Conference on Computational Intelligence, Funchal, Madeira, Portugal, 5–7 October 2009
29. Hung, K.C., Tuan, H.W.: Prioritising emergency bridgeworks assessment under military consideration using an enhanced fuzzy weighted average approach. Defence Sci. J. **60**(4), 451–461 (2010)
30. Jiang, F., Song, L., Fan, Y., Sun, Y.: Applying fuzzy weighted average approach for the selection of the emergency supplies storage location. J. Comput. Inf. Syst. **9**(10), 4101–4110 (2013)
31. Yue, L., Sun, M., Shao, Z.: The probabilistic hesitant fuzzy weighted average operators and their application in strategic decision making. J. Inf. Comput. Sci. **10**(12), 3841–3848 (2013)
32. Ngai, E.W.T., Wat, F.K.T.: Fuzzy decision support system for risk analysis in e-commerce development. Decis. Support Syst. **40**, 235–255 (2005)
33. Dubey, S.K., Gulati, A., Rana, A.: Usability evaluation of software systems using fuzzy multi-criteria approach. Int. J. Comput. Sci. Issues **9**(3)2, 404–409 (2012)
34. Maps of World. http://www.mapsofworld.com/world-map-image.html
35. The SolarCoin Foundation. http://solarcoin.org/wp-content/uploads/2014/03/SolarGIS-Solar-map-World-map-en.png
36. Eidgenössische Technische Hochschule Zürich. http://www.ethlife.ethz.ch/archive_articles/30312_global_grid/index_EN
37. Maps of World, Road Maps. http://www.mapsofworld.com/road-maps
38. World Port Source. http://www.worldportsource.com/index.php

Experimental FOWA Aggregated Location Selection Model for VLCPVPPs in MENA Region in the Very Early Engineering Design

Burak Omer Saracoglu$^{(\boxtimes)}$

Orhantepe Mahallesi, Tekel Caddesi, Istanbul, Turkey
burakomersaracoglu@hotmail.com

Abstract. Nowadays, solar energy can be used for water heating, heating, cooling and ventilation, cooking, and electricity generation. There are several ways in the research and development stages and in the operational stages to convert the sunlight into the electricity, such as using the concentrated photovoltaics (CPV). When the electricity generation technology differs, requirements, requisites, and obligations differ very much. Nowadays, the Middle East and North Africa (MENA) Region has some unique properties, especially regarding the actual conditions and situations of the social conflicts, violence, chaos, disputes and wars. Under these conditions, some of the factors loss their importance and meaning, so that they become absurd, worthless and useless. This paper presents an experimental ordered fuzzy weighted average (fuzzy OWA: FOWA) aggregated location selection model for the very large concentrated photovoltaic power plants (VLCPVPPs) in the MENA Region in the very early engineering design process stages.

Keywords: Fuzzy logic · CPV · Concentrated photovoltaic power
Design · Electricity · Energy · Engineering · Engineering design process
FuzzME · Fuzzy models of multiple-criteria evaluation
Fuzzy ordered weighted average operator · Fuzzy OWA · FOWA
MENA · Middle east and north Africa · Multiple-criteria fuzzy evaluation
North Africa · Solar power · Power · Solar
Very large concentrated photovoltaic power plants

1 Introduction

Research and development (R&D) engineers and engineers in the daily application life manage to design, build and operate the systems to use solar energy for water heating, heating, cooling and ventilation, water treatment, process heating, cooking, and electricity generation. There are several solar power technologies in the R&D stages and in the operational stages. The operational power plants based on the most commonly known solar power technologies are grouped under photovoltaics (PV) technology [visit 1], concentrated solar power (some other common terminology: concentrating solar power, concentrated solar thermal) technology (CSP) [visit 2], and concentrated photovoltaics technology (CPV) [visit 3] (see Fig. 1 for CPV power plants).

© Springer Nature Switzerland AG 2019
B. K. Ane et al. (Eds.): WSC 2014, AISC 864, pp. 36–45, 2019.
https://doi.org/10.1007/978-3-030-00612-9_4

Soitec Concentrix CPV facility   Soitec Concentrix 2 MW CPV facility

Fig. 1. Two CPV power plants (Source: [4, 5] Credit: Soitec Concentrix http://www.soitec. com/en).

Design approaches and aspects in different engineering design process stages of these technologies differ very much from each other. When an electricity generation technology differs in the electricity generation sector, the requirements, the requisites, and the obligations diverge also very much, so that the models and the analysis should be different from each other. The researchers, academics, experts and specialists simply and openly point out that the world regions have some different properties, especially for the consideration of peace versus violence. The Middle East and North Africa (MENA) Region (Algeria, Bahrain, Egypt, Iran, Iraq, etc.) is one of them (other regions of the world: the Caucasus, the Central Asia, etc.). While researchers and analysts record, track and analyze the news, the reports and the research studies by the help of some special tools and methods, they conclude that the MENA Region has some unique properties regarding the actual conditions and situations of the social conflicts, the social violence, the social chaos, the social disputes and the wars continuing for more than a decade. The countries such as Egypt, Iraq, Lebanon, Libya, Syria, and Turkey can be easily given as the examples for these discriminating situations. Under the conditions related to these issues, some of the factors loss their importance and meaning, so that they become almost absurd, worthless and useless. Henceforth, when the proposed models are built upon these worthless and useless factors, the decision makers and the analysts will spend their very valuable time to find, gather, eliminate and confirm the data and information for these factors, to evaluate these factors for each alternative and to make the computations and calculations. Some of the very interesting and informative issues have been clearly presented and shared in the War for the Greater Middle East Course by Professor Dr. Andrew J. Bacevich (Chair of International Relations; Professor of International Relations and History) [6, 7]. Therefore, it will be wise and interesting to model and analyze for an exceptional and special power plant technology and an extraordinary and outstanding power plant size like the VLCPVPPs on an engineering point of view by its own unique characteristic models specifically for the MENA Region by keeping in mind the same old situation, "war to end war", "war follows war", and "Peace to end Peace" (see [8] for the phrases). The VLCPVPPs are defined in this study as the CPV power plants that have the installed power of 1.000 MW_p (peak power) or more. The main challenge in the location selection of the VLCPVPPs in the MENA Region in the very early engineering design process stages problem from the perspective of soft computing is building up highly agreed upon, robust, manageable and reasonable models. In the current study, an experimental fuzzy OWA aggregated model for the solution of this problem is presented for helping the research and practical studies.

This paper has four sections. The following section is for the review of literature. Section 3 presents the experimental proposed ordered fuzzy weighted average (fuzzy OWA) aggregated model and the experimental application. The concluding remarks and further research studies are presented in Sect. 4.

2   Literature Review

A detailed literature review on some of the academic publication online database and journals was completed for the key terms and phrases, that could be predominantly preferred and used in the research studies. The academic publication online database and journals in this study were the ACM Digital Library [9], the Directory of Open Access Journals [10], the Google Scholar [11], the Science Direct [12], and the Springer [13]. The key terms and phrases of the current study were *"concentrated photovoltaic"* and *"fuzzy ordered weighted average"* (A), *"concentrated photovoltaic"* and *"fuzzy OWA"* (B), *"concentrated photovoltaic"* and *"FuzzME"* (C), *"concentrated photovoltaic"* and *"Fuzzy Models of Multiple Criteria Evaluation"* (D), *"very large concentrated photovoltaic solar power plant"* and *"fuzzy ordered weighted average"* (E), *"very large concentrated photovoltaic solar power plant"* and *"fuzzy OWA"* (F), *"very large concentrated photovoltaic solar power plant"* and *"FuzzME"* (G), *"very large concentrated photovoltaic solar power plant"* and *"Fuzzy Models of Multiple Criteria Evaluation"* (H), *"location"* and *"fuzzy ordered weighted average"* (I), *"location"* and *"fuzzy OWA"* (J), *"location"* and *"FuzzME"* (K), and finally *"location"* and *"Fuzzy Models of Multiple Criteria Evaluation"* (L) as shown in Fig. 2. There were 82 studies found during the literature review, however, most of the found documents were irrelevant with the topic of the current study (see Fig. 2). There were only 5 relevant previous studies of the current study. 3 closest ones were summarized in this section.

Chen and Paydar evaluated the potential irrigation expansion pasture by help of the fuzzy linguistic OWA (FLOWA) model and Analytical Hierarchy Process (AHP) by the factors of susceptibility to water logging, depth to water table, water holding capacity, soil salinity, groundwater salinity, susceptibility to acidity, inherent fertility, steepness, surface rockiness, surface texture, surface condition, subsoil structure, boron toxicity, sodium toxicity, root zone depth in the Limestone Coast of South Australia on the ArcGIS software [14]. Oh et al. selected the best anchorage area for the marine vessels (ships, etc.), by 3 main factors (ship steering, marine environment, vessel traffic), 7 sub-attributes (safe anchorage, ship steering, weather condition, geological features, sounding, proximity to sea lane, congestion), and 4 alternatives with help of the fuzzy OWA and the AHP [15]. Oh et al. solved anchorage area problem of the Ulsan Port by a fuzzy OWA operators and fuzzy AHP approach [16].

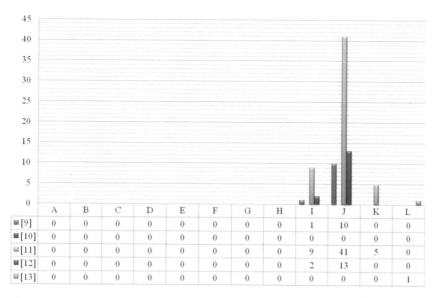

The chart contains the following data table:

	A	B	C	D	E	F	G	H	I	J	K	L
[9]	0	0	0	0	0	0	0	0	1	10	0	0
[10]	0	0	0	0	0	0	0	0	0	0	0	0
[11]	0	0	0	0	0	0	0	0	9	41	5	0
[12]	0	0	0	0	0	0	0	0	2	13	0	0
[13]	0	0	0	0	0	0	0	0	0	0	0	1

Fig. 2. Literature review summary of this study (final date of the review: 07th October 2014).

Although FOWA models were applied in some location problems, it was understood that FOWA based models had not most probably been applied in any VLCPVPPs studies (in a broader scope CPV) yet and none of the studies had been executed on the FuzzME software until 07th October 2014. As a consequence, it could be underlined that this study would be one of the first studies with the FOWA aggregated location selection model for the VLCPVPPs in the MENA Region in the very early engineering design process stages. This study would most probably be the first attempt to define the location selection problem of the VLCPVPPs, to analyze the usage possibility of the FOWA models in this problem, to find and define the factors to model and solve this problem and to show the usage of the FuzzME software (as helpful tools) in the solution of this problem.

3 The Experimental Proposed Ordered Fuzzy Weighted Average/Weight Averaging (FOWA) Aggregated Model

The current experimental proposed fuzzy OWA aggregated model is founded on some main factors and sub-factors, that are specifically selected according to the scope of this study. Unlike the other studies of the author of the Global Grid Concept and the European Supergird Concept in the 18th Online World Conference on Soft-Computing in Industrial Applications, this study aims not to take into account any social conflicts' and wars' related factors, because of very similar conditions in the MENA Region (observation period of the author: read the academic publications, news, reports and research studies from 2007 to 2014 to understand the circumstances). In this period, the MENA Region seems to be a whole war zone and a whole conflict zone of the world, where the war games have been playing by several armed forces and groups acting like several console game players. In other words, all in war and all in peace conditions have been occurred in the MENA Region for years and years. Thus, the Direct Normal Irradiance (DNI) (Factor 1: F_1), the HVDC & HVAC electricity grid infrastructure (F_2) (HVDC: High-Voltage Direct Current, HVAC: High-Voltage Alternating Current), with its two sub-factors as the HVDC transmission system network (F_{21}) and the HVAC transmission system network (F_{22}), the natural disaster (F_3), with its three sub-factors as the earthquakes (F_{31}), the floods (F_{32}), and the volcanic activities (F_{33}), and finally the transportation (F_4), with its five sub-factors as the air transportation (F_{41}), the railway transportation (F_{42}), the road transportation (F_{43}), the inland water way transportation (F_{44}), and the ocean-sea transportation (F_{45}) are taken into account in this study.

One of the important and necessary factors in this model is the HVDC & HVAC electricity grid infrastructure. The distance and the conditions between the locations of the VLCPVPPs and the HVDC & HVAC transmission grid should be evaluated by this factor. For instance, the HVDC systems between Italy and Algeria, Italy and Tunisia, Turkey and Egypt, Turkey and Syria, and finally Egypt-Jordan-Syria-Turkey have been investigating for a long while (visit [17–19]). It can be realized with ease that, the factors related with the social conflicts and wars do not have any major influences in the countries such as Egypt, Jordan, Syria, and Turkey in the MENA Region on these days, because all in war condition sticks to the evaluations for all of them, so that these kinds of factors are eliminated in this model.

The current study was directly modeled on the FuzzME Software, that was developed and presented by Holecek, Talasova, Pavlacka and Bebcakova (see [20, 21] for FuzzME). Moreover, during the design and the modeling of the linguistic terms, the scales, and the hedges of the factors, the research studies and the findings of Likert (1903–1981) (5 Likert scales) [22], Zadeh (1921–alive) (fuzzy set, fuzzy logic, and hedges) [23], Miller (1920–2012) (magical number 7) [24], Shiffrin (1968–alive) and Nosofsky (alive) (magical number 7, 7 ± 2 rule) [25], and finally Rohrmann (alive) [26] were tried to be adopted and used in a very organized manner (see Table 1).

In this study, the aggregation of the partial evaluations were performed with the ordered weight averaging (OWA) operator, which had been proposed by Yager (alive) (contributions in the computational intelligence and the decision making), based on the evaluation tools of the FuzzME. The principles, and the properties of the fuzzy OWA application in the current study was pillared on the previous studies in this subject [27–30]. Yager explained and expressed very clearly that *"It is important to emphasize the fact that the weights, the W_i's, are associated with a particular ordered position rather than a particular element. That is W_i is the weight associated with the ith largest element whichever component it is."* [28, p. 185]. Holecek and Talasova explained the aggregation of the partial evaluations by FOWA on the FuzzME as *"Again, the ordered fuzzy weighted average requires that the goal corresponding with the given node is decomposed into disjunctive goals of the lower level. In contrast to the fuzzy weighted average, the usage of this aggregation operator supposes special user's requirements concerning the structure of partial fuzzy evaluations."* [31, p. 43].

Table 1. The representative normalized membership functions on the FuzzME (open the presentation file and the FuzzME model file).

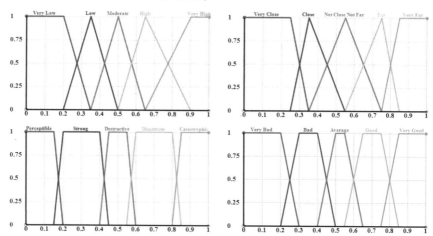

* Factors: DNI F_1 (up left) (Increasing Scale: higher values are better), HVDC & HVAC F_2 (up right) (Decreasing Scale: lower values are better), Natural Disaster F_3 (down left) (Decreasing Scale: lower values are better), Transportation F_4 (down right) (Increasing Scale: higher values are better).
* Note: Increasing Scale: values in the real life are normalized with the maximum value, Decreasing Scale: first, values in the real life are normalized with the maximum value; second, normalized values are calculated and changed as x to (1-x) or directly decreasing scale selected on the FuzzME.

The approaches and importance of the assignments of the weights were underlined by Yager (also by the other researchers) in his publications and one of them was shortly extracted as *"A second approach is to try to give some semantics or meaning to the W_i's. Then based upon these semantics we can have the decision maker directly provide the values for the W_i's. This approach also will provide some further insight into the meaning of the OWA operators we have just introduced."* [28, p. 186] (see also [32]). In this experimental model, this approach was tried to be applied to all of the

Table 2. The weights of fuzzy OWA on the FuzzME (open the presentation file and the FuzzME model file).

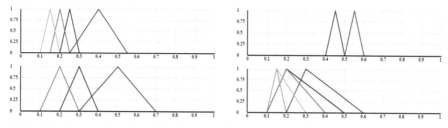

* Note: Selection (up left), F_2 (up right), F_3 (down left), F_4 (down right), F_1 (no sub-factor).

weight assignments on the FuzzME (see Table 2), by first assigning the uniform weights and then defining the fuzzy weights guided with the warning messages.

The experimental application in this study was executed on a few randomly selected provinces and districts in the MENA regions by using the FuzzME Software. The FuzzME Software's capability and flexibility were tried to be used as much as possible for the evaluations of the alternatives. Some of the criterion value was selected in either simple or extended scale type (see Table 3). The provinces & districts of the countries in this study were Libya-Sirte, Libya-Jufra, Libya-Murzuq, Egypt-New Valley, Egypt-Red Sea, and Egypt-Suez. These options were not presented directly on the current study and the FuzzME software, because of the aim and the intention of this study. The evaluations were made by only one decision maker and very limited data and information had been gathered, understood, perceived, internalized and used based on both some official documents and web pages, and unofficial documents and websites (see Fig. 3).

Table 3. The experimental alternatives on the FuzzME (open the presentation file and the FuzzME model file).

Abbreviation	F_1	F_{21} & F_{22}	F_{31} & F_{32} & F_{33}	F_{41} & F_{42} & F_{43} & F_{44}
Alt1	Simple	Extended	Extended & Simple	Extended & Simple
Alt2	Simple	Extended	Extended & Simple	Extended & Simple
Alt3	Simple	Extended	Extended & Simple	Extended & Simple
Alt4	Simple	Extended	Extended & Simple	Extended & Simple
Alt5	Simple	Extended	Extended & Simple	Extended & Simple
Alt6	Simple	Extended	Extended & Simple	Extended & Simple

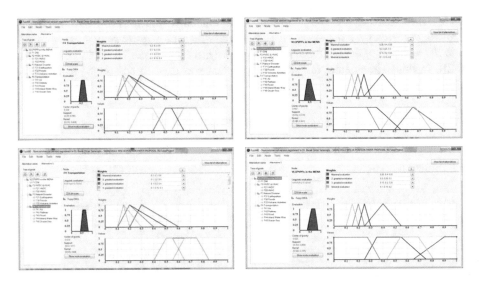

Fig. 3. Screenviews of alternative 1 & alternative 5 on the FuzzME (open the presentation file and the FuzzME model file).

The alternatives were directly evaluated one by one on the FuzzME, and as soon as all of the evaluations were finalized the alternatives were directly sorted on the FuzzME (sort by evaluation tab) as the Alternative 5, the Alternative 6, the Alternative 3, the Alternative 4, the Alternative 2, and the Alternative 1 in this experimental application (see Fig. 4). Based on this finding, in the practical life, the concentration and the efforts for selecting the most satisfying and advisable location of the very large concentrated photovoltaic power plants in the MENA Region would be directed to the Alternative 5 and then the Alternative 6.

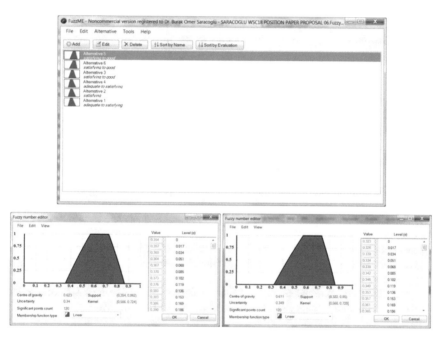

Fig. 4. List of alternatives for sort by evaluation on the FuzzME (open the presentation file and the FuzzME model file) * Note: All (up), Alternative 5 (down left), Alternative 6 (down right).

4   Conclusions and Future Work

It is hoped that the current study presents the usability and adaptability of the ordered weight averaging (OWA) operator with the fuzzy set and logic combination to build the fuzzy OWA applications for the solution of the location selection problem for the very large concentrated photovoltaic power plants (VLCPVPPs) in the Middle East and North Africa Region in the very early engineering design process stages by the help of FuzzME. It is believed that the more sophisticated and developed models of this one shall help the electricity generation sector (in this case CPV) to start to investigate the possibility of the engineering, the construction and the operation of the VLCPVPPs in the MENA Region (also in other regions of our world). The following studies should first focus on the identification and the definition of the location selection factors of the VLCPVPPs and the utility-scale CPV power stations, afterwards the fuzzification of these factors should specifically be studied in detail, in addition the weights of the fuzzy OWA operators should carefully be studied and finally the real world applications should be conducted to help the international organizations, the multinational foundations, the governments and the investors.

Acknowledgments. The author would like to thank to Dr. Bernadetta Kwintiana Ane (conference), Dr. Ronald Robert Yager (publications), and Dr. Pavel Holeček (FuzzME). This study shall never be finalized and submitted to the conference without their consideration, guidance, and help. Please send your comments, feedbacks and criticisms to my e-mail (burakomersaracoglu@hotmail.com) in any format at any time. Your feedback will be very important and valuable for me during the development process of the models and systems for the real life applications.

References

1. Wikipedia: Wikimedia Foundation Inc., Photovoltaics. http://en.wikipedia.org/wiki/Photovoltaics
2. Wikipedia: Wikimedia Foundation Inc., Concentrated solar power. http://en.wikipedia.org/wiki/Concentrated_solar_power
3. Wikipedia: Wikimedia Foundation Inc., Concentrated photovoltaics. http://en.wikipedia.org/wiki/Concentrated_photovoltaics
4. Soitec CPV Installations. http://www.soitec.com/en/products-and-services/solar-cpv/our-references
5. RenewableEnergyWorld.com: Concentrix Reveals CPV Plant Efficiency Figures. http://www.renewableenergyworld.com/rea/news/article/2009/06/concentrator-pv-plant-efficiency-figures-revealed
6. Boston University Frederick S: Pardee School of Global Studies, Andrew J. Bacevich. http://www.bu.edu/pardeeschool/academics/faculty/faculty-emeriti/bacevich
7. Bacevich, A.J.: BUx: INTL301x War for the Greater Middle East Course, Module 1 & 2, EdX online courses from MITx, HarvardX, BerkeleyX, UTx. https://www.edx.org
8. Pagden, A.: Worlds at War: The 2,500-year Struggle between East and West. Oxford University Press, Oxford (2008)
9. ACM Digital Library. http://dl.acm.org
10. Directory of Open Access Journals. http://doaj.org
11. Google Scholar. http://scholar.google.com.tr
12. Science Direct. http://www.sciencedirect.com
13. Springer. http://www.springer.com/?SGWID=5-102-0-0-0
14. Chen, Y., Paydar, Z.: Evaluation of potential irrigation expansion using a spatial fuzzy multi-criteria decision framework. Environ. Model Softw. **38**, 147–157 (2012)
15. Oh, S.W., Park, G.K., Lee, C.Y.: Integration of decision inputs with OWA operators for MCDM problems, ISIS (International Conference on Soft Computing and Intelligent Systems) 2007. In: Proceedings of the 8th Symposium on Advanced Intelligent Systems 2007, vol. 9, pp. 252–257 (2007)
16. Oh, S.W., Park, G.K., Park, J.M., Suh, S.H.: A study on the location analysis using spatial analysis and ordered weighted averaging operator weighting functions. J. Adv. Comput. Intell. Intell. Inf. **12**(6), 529–536 (2008)

17. The European Union Neighbourhood Info Centre. http://www.enpi-info.eu/index.php
18. The Paving the Way for the Mediterranean Solar Plan Project. http://www.pavingtheway-msp.eu/index.php
19. Green Design & Building Professionals, Solaripedia. http://www.solaripedia.com
20. Talasova, J., Holecek, P.: Multiple-criteria fuzzy evaluation: the FuzzME software package. In: IFSA/EUSFLAT Conference 2009, pp. 681–686 (2009)
21. The website of FuzzME. http://www.fuzzme.net. http://fuzzme.wz.cz
22. Likert, R.: A Technique for the Measurement of Attitudes, vol. 140. Archives of Psychology, New York (1932)
23. Zadeh, L.A.: A fuzzy-set-theoretic interpretation of linguistic hedges. J. Cybern. $2(3)$, 4–34 (1972)
24. Miller, G.A.: The magical number seven, plus or minus two: some limits on our capacity for processing information. Psychol. Rev. 63, 81–97 (1956)
25. Shiffrin, R.M., Nosofsky, R.M.: Seven plus or minus two: a commentary on capacity limitations. Psychol. Rev. $101(2)$, 357–361 (1994)
26. Rohrmann, B.: Verbal qualifiers for rating scales: sociolinguistic considerations and psychometric data. Project report, University of Melbourne, Australia (2007). http://www.rohrmannresearch.net/pdfs/rohrmann-vqs-report.pdf
27. Stoklasa, J., Talasova, J., Holecek, P.: Academic staff performance evaluation – variants of models. Acta Polytechnica Hungarica $8(3)$, 91–111 (2011)
28. Yager, R.R.: On ordered weighted averaging aggregation operators in multicriteria decision making. IEEE Trans. Syst. Man Cybern. $3(1)$, 183–190 (1988)
29. Zhou, S.M., John, R.I., Chiclana, F., Garibaldi, J.M.: On aggregating uncertain information by type-2 OWA operators for soft decision making. Int. J. Intell. Syst. $25(6)$, 540–558 (2010)
30. Herrera, F., Herrera-Viedma, E.: Linguistic decision analysis: steps for solving decision problems under linguistic information. Fuzzy Sets Syst. 115, 67–82 (2000)
31. Holecek, P., Talasova, J.: Fuzzme: a new software for multiple-criteria fuzzy evaluation. Acta Universitatis Matthiae Belii Ser. Math. 16, 35–51 (2010)
32. Yager, R.R.: Quantifier guided aggregation using OWA operators. Int. J. Intell. Syst. $11(1)$, 49–73 (1996)

Fuzzy Demand Forecast Classification and Fuzzy Pattern Recognition for Distributed Production

Dieter Roller$^{(\boxtimes)}$ and Erik Engesser

Institute of Computer-Aided Product Development Systems, Universität
Stuttgart, Universitätsstraße 38, 70569 Stuttgart, Germany
{Dieter.Roller,
Erik.Engesser}@informatik.uni-stuttgart.de

Abstract. Global Corporations have to manage distributed production over the whole world. Therefore global supply chains are needed. This paper discusses the problem how global production plants and their supply chains can be classified. The classification focuses on demand and supply of production and supply chain. The problem of forecasting the demand of a global supply chain is introduced. Objective of the paper is to show solutions of the mentioned problems by using fuzzy classification and fuzzy pattern recognition methods. The approach is to use the classification methods fuzzy c-means (FCM) and Improved Fuzzy Clustering (IFC). Supply and demand patterns can be found with fuzzy pattern recognition. Therefore the methods Multi Feature Pattern Recognition and Fuzzy Inference System Type-2 (FIS 2) with neural network methods are introduced. The solution of the mentioned approach is realized by the application PROCAS (Process Optimization, Control, Analysis and Simulation). PROCAS uses a data warehouse database for multidimensional fuzzy classification data and Business Intelligence (BI) functionalities. The key result is that fuzzy classification and fuzzy pattern recognition applications improve the planning and operating of supply and demand in a distributed production and a global supply chain.

Keywords: Fuzzy classification · Fuzzy pattern recognition
Distributed production · Global supply chain · FCM · IFC · FIS2
Business Intelligence · Data warehouse · Multidimensional fuzzy classification

1 Introduction

Global corporations have to manage distributed production over the whole world. They build production plants in international locations. The strategy is to produce local products for local markets. Especially new growing markets like China require a local production in China. Reasons therefore are for example legislation, market presence etc. The process of economics in such growing markets shows that well-known corporations create production plants at upcoming and important market locations.

Global supply chains grow to a worldwide network. By reason of distributed production global supply chain are generated. The demand of products leads to demands of

B. K. Ane et al. (Eds.): WSC 2014, AISC 864, pp. 46–55, 2019.
https://doi.org/10.1007/978-3-030-00612-9_5

components and parts. Often the in-house production depth is low. This causes a global supplier network with multiple tiers. Sometimes suppliers of low cost countries are chosen to reduce costs or suppliers of high technological countries are chosen to get technical components like microcontrollers or electrical devices. A seat of a high class automobile for example needs leather from Africa, metal components from East Europe and electronic components from Asia. With these components the seats are built nearby any production line. The consequence is that all the parts have to be supplied to the distributed production plants in dependency of the demand of the local markets.

This paper discusses the problem how global production plants and their supply chains can be classified. The problem of classification global production is that multiple classification criteria for different kinds of markets, production plants and supply chain exist. Classification criteria are for example order volume, product variants, part volume, part variance, seasonal variation and so on. A good quality of demand forecasting is reached when the forecast classification of demand trend, demand level and seasonal factor agrees to the happened demand classification.

Further classification criteria like precision and uncertainty of demand forecasts are discussed in this paper. The problem is how uncertainty of market behavior can be included and what precision can be expected. Forecasting of demand of a global supply chain is difficult because of uncertainties of global market behaviour. Demands of global markets are influenced by multiple criteria. Local situations of each markets like income, reactions on advertising, politics, etc. are examples of such criteria. An example of distributed markets, production and supply chain is shown in Fig. 1.

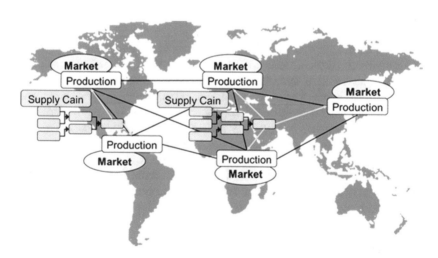

Fig. 1. Distributed supply chain and production

The objective of the paper is to show methods and an application for planning and controlling distributed productions and global supply chains. These methods are explained and an application shows how the realized methods work.

The methods support forecasting the demand of products for global markets. Classifications of demand and supply situations can be made. The classification of this situation helps to analyze the current state. These analyze support decision findings for reacting on volatile markets or correcting supply capacities.

The methods find regularities in demand and supply in distributed production. Regularities can for example be seasoning behavior or advertising effects. The recognition of regular demand patterns help to predict the demand behavior. A precise demand prediction helps the suppliers to produce the correct amount of parts.

Pattern recognition is also used make observations of demand and supply behavior. Importance of these observations is to find supply bottlenecks in an early state. If a supply bottleneck is found for specific parts capacities can be adjusted. Another alternative is to find alternative parts or other suppliers.

The introduced methods makes estimations and calculations of uncertainty and imprecision of demand and supply. Therefore using fuzzy logic theory is used. Existing fuzzy classification and clustering techniques are used and extended. These fuzzy technologies are realized in an application which is introduced.

Aggregated planning of part variants in distributed production can be done with the extended fuzzy technologies in the application. The demand of parts for distributed plans can be aggregated for several products which consist of an amount of parts.

Outline of the paper is: in Sect. 2 the proposed approach and model is described, the result is explained in Sect. 3, finally in Sect. 4 is the conclusion.

2 Proposed Approach and Model

The approach of this paper is to solve the problem by fuzzy classification and fuzzy pattern recognition. Fuzzy classification methods are used for demand classification. Type-2 fuzzy classification and clustering techniques are used to achieve the named objects. The methods FCM and IFC are introduced and applied to make demand and supply classification. The demand and supply criteria were classified in structures by FCM. A further approach is to use fuzzy pattern recognition methods to find demand and supply patterns. Multi feature pattern recognition is applied to find patterns for n-dimensional criteria. FIS 2 with neural network methods are used to find and learn patterns of demand and supply behavior. These patterns are used to predict the demand variability which is used for demand and supply synchronization. The paper introduces a realization and application of the fuzzy methods. The application is called PROCAS (Process Optimization, Control, Analysis and Simulation). PROCAS uses a data warehouse. The application consist of Business Intelligence (BI) techniques data extraction, multidimensional fuzzy classification, data mining, OLAP reports and charts technologies. The database has a dimensional modeling structure with dimensional tables and fact tables. PROCAS supports BI-functionalities to analyze classification results. Pattern recognition of demand and supply are calculated by PROCAS. The functionality of aggregated planning of demand over distributed production plants is shown. Additional the functionality demand prediction and demand supply synchronization is represented.

Zadeh [ZA09] demonstrates extended fuzzy logic as an important capability to reason precisely with imperfect, imprecise, uncertain and incomplete information. In [ZA05] he demonstrates the approach form imprecise to granular probabilities. Transformation on binary survival aggregation functions are proposed by de Baets et al. in [BA12]. Aggregation of monotone reciprocal relations for group decision making is introduced in [RA11]. Multi-class classification problems with partial class memberships are discussed in article [WA11]. Gradualness, uncertainty and bipolarity of fuzzy sets are discussed by Dubois and Prade in article [DU12]. Hüllermeier presents in [HÜ05] fuzzy methods in machine learning and data mining. In study [PE12] of Pedrycz, Hirota and Dong a concept of granular representation and granular computing with fuzzy sets is shown. A new fuzzy radial basis function-based polynomial neural network is demonstrated in [RO11]. In article [BA11] the use of a fuzzy granulation and degranulation criterion for assessing cluster validity is introduced. Reformat, Pedrycz and Pizzi introduce in [RE04] a development method of granular based models represented by fuzzy neural networks. Runkler and Bezdek represent in [RU99] function approximation with polynomial membership functions and alternating cluster estimation. Clustering documents with labeled and unlabeled documents using fuzzy semi-Kmeans are explained by Liu et al. [LI13].

In our approach the methodology is about demand prediction of distributed production and global supply chain.

2.1 Demand Prediction

Forecasts of future demand are essential for making supply chain decisions. Figure 2 is an example of demand over season is shown.

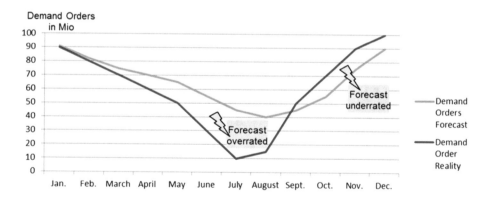

Fig. 2. Demand orders forecast and reality

The green line shows the demand orders forecast. The red line is the demand order in reality. In this example from February to September the forecast was overrated. From September to December the forecast was underrated. A good demand forecast is when overrates and underrates are as small as possible. In this paper "Time Series Forecasting Methods" is used. The systematic components are level, trend and seasonal factor. In the static method it is estimated that level, trend and seasonality do not vary within the systematic component. The systematic component is calculated by (1).

$$D_{Observed} = D_{Syst} + D_{Uncert} \tag{1}$$

where $D_{Observed}$ = Observed Demand, D_{Syst} = Syst. Demand, and D_{Uncert} = Uncert. Demand.

In Fig. 3 an observed demand over season is displayed. In this example the observed demand consist of many demand deviations:

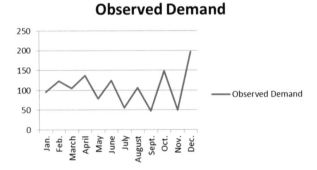

Fig. 3. Observed demand over season

2.2   Demand Prediction Uncertainty and Uncertainty Error

In (1) the observed demand consist of systematic demand and uncertainty demand. The uncertainty demand is calculated by (2). It is the product of the systematic demand and the uncertainty error factor E_{Uncert}. The uncertainty error factor is defined as exponential function in (3) for time periods T_{Period}. E_{UExp}. is the uncertainty exponential error. The observed demand vector is calculated by (4). The observed best case and the observed worst case demand describes the demand forecast area.

$$D_{Uncert} = D_{Syst} * E_{Uncert} \tag{2}$$

$$E_{Uncert} = \pm E_{Uconst} * e^{E_{UExp}*T_{Period}} * T_{Period} \tag{3}$$

$$\vec{D}_{Observed} = \vec{D}_{Syst} * (1 + E_{Uncert}) \tag{4}$$

In Fig. 4 the observed demand with uncertainty demand and worst case is shown. In this case the Uncertainty Exponential Error E_{UExp} is 0 and the uncertainty rises linear.

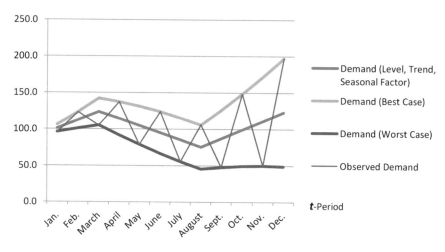

Fig. 4. Observed demand with linear uncertainty demand

In Fig. 5 an exponential uncertainty demand is shown. If there is a high uncertainty demand the Uncertainty Exponential Error E_{UExp} factor has to be used to describe the uncertainty demand area.

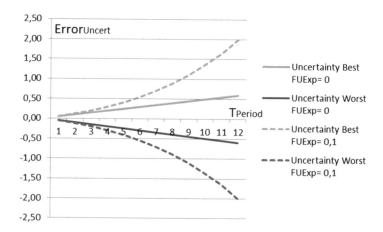

Fig. 5. Observed demand with exponential uncertainty demand

2.3   Fuzzy Sets and Fuzzy Membership Function

The fuzzy set for planned demand forecasting is defined in following fuzzy sets: Poor, Fair, Good, Very good, excellent. In Fig. 6 the fuzzy membership function type 2 is shown:

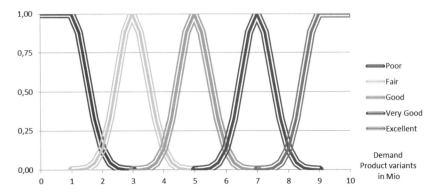

Fig. 6. Fuzzy membership function type 2 for clustering

2.4   Business Intelligence

PROCAS is a novel research tool. Existing and innovative process designs for distributed production and global supply chains can be modeled, simulated and optimized. Planned, simulated or reality data can be analyzed by fuzzy classification and fuzzy pattern recognition.

PROCAS simulates the production and supply chain process models. The results were stored in a data warehouse database. With these results PROCAS calculates the fuzzy classification and fuzzy pattern recognition. The architecture of the data warehouse is represented in Fig. 7:

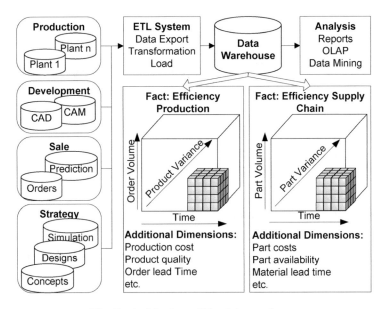

Fig. 7. Architecture of the data warehouse

The data warehouse is filled by further databases of production, development, sales and strategy. The data is loaded by data export in to the data warehouse. The data is stored in a multi-dimensional structured database. Dimensional modelling is a logical design technique for structuring data. The database consists of fact tables and dimensional tables.

Fact tables correspond to business process measurements events. Figure 7 shows two multidimensional facts "Production" and "Supply". Dimensions of fact production are e.g. time, order volume, production cost. The fact supply chain consist of time, part volume, part cost etc. The dimensional data are used for fuzzy classification and fuzzy pattern matching. The results of fuzzy classification and fuzzy pattern matching are used for fuzzy decision making and efficiency measuring.

3   Results

3.1   Demand of Product Variants

The pattern recognition method calculates the precision between the forecast and the reality demand. The application supports error calculation and determines the uncertainty of the forecast. The demand of orders consists of various product variants. In Fig. 8 a fuzzy set of product variants are shown.

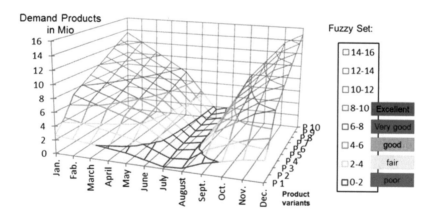

Fig. 8. Demand product variants over season

3.2   Fuzzy Data Mining

PROCAS supports the aggregated planning of part demand with the method "Multi Feature Pattern Recognition". The pattern features are dimensions of Fact table Supply Chain in the data warehouse. Pattern features are: Time, part demand, part variant distributed plants and global supply chains. In the n-dimensional data cube the dimensions can be aggregated to get demand patterns. The aggregation of product variants in Fig. 9 shows the demand pattern of products for the distributed plants.

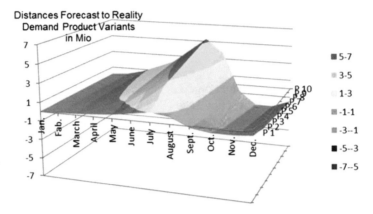

Fig. 9. Pattern recognition forecast and reality distance

The demand is classified in Poor, Fair, Good, Very good, excellent and Maximal capacity exceeded. Figure 9 shows aggregated demands of distributed plants:

3.3 Fuzzy Pattern Recognition

The application supports error calculation and determines the uncertainty of the forecast. The concept of PROCAS is to use fuzzy pattern recognition. In this way forecast patterns of pattern sets can be evaluated by training with reality demand patterns. Therefore the overall distance between the forecast patterns and the reality patterns has to be minimized. An example of pattern recognition is shown in Table 1.

Table 1. Fuzzy pattern recognition

Forecast Demand	Distance to Reality	Distance Pattern
Pattern 1	62	
Pattern 2	163	
Pattern 3	242	

4 Conclusion

Fuzzy classification methods are successful approaches to classify demand and supply. Uncertainty can be measured by an uncertainty error. The error is defined by an area of best case and worst case. With fuzzy sets these fuzzy error can be classified. Fuzzy pattern recognition improves demand forecasting by calculating forecast and reality demand patterns. The quality of demand patterns can be measured by the fuzzy error area. The application PROCAS supports demand forecasting with these fuzzy technics.

Future works are to implement the fuzzy classification technics and pattern recognition in more detail and calculate much more patterns. Further steps are to examine and classifying aggregated planning of demand and supply and demand synchronization for distributed production and global supply chain.

References

[ZA09]   Zadeh, A.: Toward extended fuzzy logic—a first step. Fuzzy Sets Syst. **160**, 3175–3181 (2009)

[ZA05]   Zadeh, A.: From imprecise to granular probabilities. Fuzzy Sets Syst. **154**, 370–374 (2005)

[BA12]   De Baets, B., De Meyer, H., Mesiar, R.: Binary survival aggregation functions. Fuzzy Sets Syst. **191**, 83–102 (2012)

[RA11]   Rademaker, M., De Baets, B.: Aggregation of monotone reciprocal relations with application to group decision making. Fuzzy Sets Syst. **184**, 29–51 (2011)

[WA11]   Waegeman, W., Verwaeren, J., Slabbinck, B., De Baets, B.: Supervised learning algorithms for multi-class classification problems with partial class memberships. Fuzzy Sets Syst. **184**, 106–125 (2011)

[DU12]   Dubois, D., Prade, H.: Gradualness, uncertainty and bipolarity: making sense of fuzzy sets. Fuzzy Sets Syst. **192**, 3–24 (2012)

[HÜ05]   Hüllermeier, E.: Fuzzy methods in machine learning and data mining: Status and prospects. Fuzzy Sets Syst. **156**, 387–406 (2005)

[PE12]   Pedrycz, A., Hirota, K., Pedrycz, W., Dong, F.: Granular representation and granular computing with fuzzy sets. Fuzzy Sets Syst. **203**, 17–32 (2012)

[RO11]   Roth, S., Oh, S., Pedrycz, W.: Design of fuzzy radial basis function-based polynomial neural networks. Fuzzy Sets Syst. **185**, 15–37 (2011)

[BA11]   Bandyopadhyay, S., Saha, S., Pedrycz, W.: Use of a fuzzy granulation–degranulation criterion for assessing cluster validity. Fuzzy Sets Syst. **170**, 22–42 (2011)

[RE04]   Reformat, M., Pedrycz, W., Pizzi, N.: Building a software experience factory using granular-based models. Fuzzy Sets Syst. **145**, 111–139 (2004)

[RU99]   Runkler, T.A., Bezdek, J.C.: Function approximation with polynomial membership functions and alternating cluster estimation. Fuzzy Sets Syst. **101**, 207–218 (1999)

[LI13]   Liu, C.-L., Chang, T.-H., Li, H.-H.: Clustering documents with labeled and unlabeled documents using fuzzy semi-Kmeans. Fuzzy Sets Syst. **221**, 48–64 (2013)

Modeling Excess Carbon Dioxide Emissions from Traffic Congestion in Urban Areas

Fengxiang Qiao[1](✉), Ling Liu[1], Wen Long[2], and Lei Yu[1]

[1] Texas Southern University, Houston, TX, USA
{qiao_fg,yu_lx}@tsu.edu, liulingsuper1983@gmail.com
[2] University of Houston, Houston, TX, USA
wlong@uh.edu

Abstract. Transportation contributes to the large amount of greenhouse gas emissions, especially for CO_2. Urban planning, transportation operation, and energy use definitely affect the amount of CO_2 emitted in an urban area. It would be very meaningful to quickly quantify the possible change of CO_2 with different planning scenarios. This paper examines and forecasts the yearly change of CO_2 emissions from traffic congestion by developing a quick response model based on a fuzzy table look-up scheme, linking the model to real-world data on city patterns and traffic conditions from 2012 Annual Urban Mobility Report by Texas Transportation Institute and United States Census Bureau. With a case study of Austin, Texas, to cap CO_2 emissions from congestion at 2.61% growth per year, there is a need to keep the population-weighted density unchanged, VMT annual growth rate capped at 2.32%, and annual excess fuel consumption growth rate capped at −1.60%.

Keywords: Congestion · Fuzzy system · Table look-up scheme
Population-weighted density · Vehicle fuel consumption
Vehicle miles traveled · Carbon dioxide emissions · Urban planning

1 Introduction

According to the United States Environmental Protection Agency (EPA), 1,745.5 Million Metric Tons carbon dioxide CO_2 comes from transportation sources, accounting for 28% of such emissions. Transportation accounts for approximately one third of the United States' CO_2 emissions inventory. The transportation sector has dominated the growth in U.S. carbon dioxide emissions since 1990, accounting for 69 percent of the total increase in U.S. energy-related carbon dioxide emissions [1]. In order to reduce future CO_2 emissions, transportation policy makers are looking for more efficient vehicles with the increased use of carbon-neutral alternative fuels.

However, less attention has been placed on CO_2 emissions reduction by released traffic congestions. CO_2 emissions can be lowered down by improving traffic operations, adjusting land use types, and modify urban planning strategies. When traffic congestion decreases, fuel consumption and CO_2 emissions will be also reduced accordingly. The key issue is to what extend an emissions reduction can achieve from reduced congestions in urban area [2].

© Springer Nature Switzerland AG 2019
B. K. Ane et al. (Eds.): WSC 2014, AISC 864, pp. 56–68, 2019.
https://doi.org/10.1007/978-3-030-00612-9_6

Texas A&M Transportation Institute (TTI) conducted an Urban Mobility Study that includes estimates of traffic congestion in many large cities and the impact on society. As traffic volume increases faster than road capacity, congestion has gotten progressively worse, despite the push toward alternative modes of transportation, new technologies, innovative land-use patterns, and demand management techniques [3]. Household travel by light-duty vehicles contributes most to road congestion and resulting CO_2 emissions, accounting for over 80% of miles traveled on U.S. roadways and 75% of the CO_2 emissions from mobile sources [4].

Kopp identifies three factors that affect CO_2 emissions from light-duty vehicles: vehicle use (typically expressed as vehicle miles traveled or VMT), fuel economy (typically expressed in miles per gallon or mpg), and net greenhouse gas (GHG) emissions associated with the production and fuel consumption from transportation sector [5].

Light-duty vehicle miles traveled (VMT) per licensed driver, reached 12,900 miles per year in 2007 and decreased to about 12,500 in 2012. This shift in travel behavior is important because it directly influences light-duty vehicle (LDV) energy demand for transportation fuels such as gasoline and diesel.

Palm points out that, to increase urban densities might explicitly impact the household's transportation options. This will make a change on the carbon dioxide of transportation sector [6]. Mahmood et al. find that, population density has the positive impacts on carbon dioxide emissions [7]. Denser cities tend to produce relatively less CO_2, as people travel shorter distance and utilize more public transportation. Thus, population density could act as an important tool in reducing carbon dioxide emissions from passenger vehicle travels.

Based on the aforementioned literature, population-weighted density, daily vehicle miles traveled, and annual excess fuel consumed due to congestion are chosen to analysis their impact on excess CO_2 emissions due to congestion. The reason why the population-weighted density rather than a traditional density measure are chosen is that, the traditional measure of density isn't very informative about the geographic distribution of population in an urban area. Dividing population by area is highly sensitive to how the area is defined. A method that overcomes this problem is to use the population-weighted density, which counts for the average number of people per square mile of land area; distances are measured to the city hall or similar municipal building of the metro area's first-named principal city [8]. The average is less impacted by large unpopulated areas, largely eliminating boundary games. Figure 1 shows the trend of fuel consumption by light duty vehicle, vehicle miles traveled from gasoline, diesel transportation fuel use, CO_2 emissions from gasoline, travel time index, and diesel transportation fuel use.

Excess CO_2 emissions due to congestion is closely related to many active research fields in transportation, including urban planning, transportation planning, traffic operation simulation, transportation network implementation and vehicle emission. A full-scale model of congestion and CO_2 emission is more time-consuming to simulate, and too many variables involved in modeling could incur extra errors. It is necessary to develop a quick response procedure with a model to forecast excess CO_2 emissions due to congestion. In the next section, the CO_2 emissions are examined and forecasted by linking the model with real-world city patterns and traffic conditions.

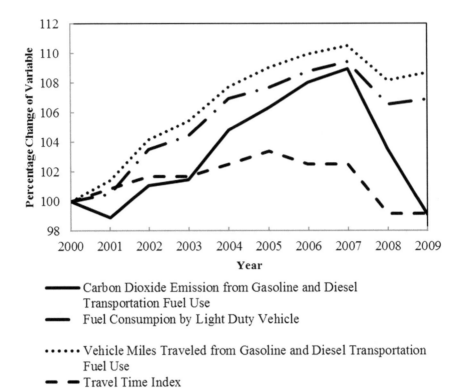

Fig. 1. 2000–2009 Percentage Change of Variables in United States while compared with 2000 Values for: (1) CO_2 emissions from gasoline and diesel transportation fuel use; (2) fuel consumption by light duty vehicle, (3) vehicle miles traveled from gasoline and diesel transportation fuel use; and (4) travel time index. (Source: [9, 10])

2 Fuzzy Modeling of Yearly Change of CO_2 Due to Congestion

2.1 Model Selection

The variables that affect excess CO_2 emissions due to congestion could be population; population density; population-weighted density; peak period travelers; number of commuters; daily vehicle miles of travel on freeway, arterial street, public transportation; vehicle miles travelled; total mileage of highways; congested travel; number of rush hours; annual hours of delay; travel time index; freeway planning time index; annual congestion cost; annual effects of operations treatment-delay reduction, delay reduction per auto commuter, additional wasted fuel, annual congestion cost increase; roadway congestion index and annual excess fuel consumed.

From the literature review (1) population-weighted density, (2) daily vehicle miles traveled, and (3) annual excess fuel consumed due to congestion are chosen to analysis the impacts on excess CO_2 emissions due to congestion. Linear models were firstly tried but the modeling errors were very big. Therefore, the nonlinear models are even proper for this case.

2.2 Methodology Description

Fuzzy logic theory is capable of dealing with systems where the input-output relationships are too complex and where the human knowledge is reachable to be combined into the system [11]. The fuzzy table look-up scheme is proven to be an effective way for nonlinear modeling [12, 13]. Therefore the fuzzy table look-up schedule is selected in this research to model the changes in CO_2 due to congestion in urban areas. Figure 2 is the model framework.

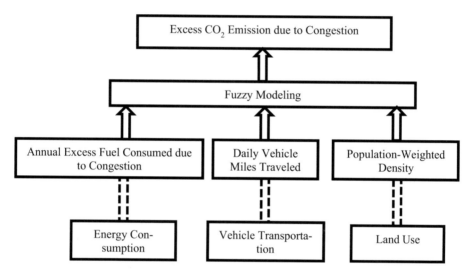

Fig. 2. The modeling framework

2.3 Fuzzy Table Look-up Scheme

The modeling of fuzzy table look-up scheme depends on collected input-output data pairs [12, 13], which are represented as

$$\left(x_1^p, x_2^p, x_3^p, \ldots x_n^p\right) \in U \tag{1}$$

Where $U = [\alpha_1, \beta_1] \times [\alpha_2, \beta_2] \times [\alpha_3, \beta_3] \times \ldots \times [\alpha_n, \beta_n]$, $p = 1, 2, 3, \ldots N$ represents each of data pair and N is the total number of data pairs. The input variables could be gender, age, education background, and driving experience. The output is value of excess CO_2 emissions due to congestion, which is defined in the following domain.

$$y_0^p \in V = \left[v_y, \omega_y\right] \subset R \tag{2}$$

The proposed fuzzy system should be based on the rule base constructed from these N input-output pairs. The following scheme can be used to design the fuzzy system.

2.4   Define the Fuzzy Sets to Cover the Input and Output

For each $[\alpha_i, \beta_i]$, $i = 1, 2\dots n$, define N_i fuzzy sets $A_i^j (j = 1, 2\dots N_i)$, which should be complete in $[\alpha_i, \beta_i]$. This means, for any $x_i \subset [\alpha_i, \beta_i]$, there exists A_i^j such that its membership $\mu_{A_i^j}(x_i) \neq 0$. The pseudo-trapezoid membership function is such a candidate.

2.5   Generate One Rule from One Input-Output Pair

First, for each input and output pair $\left(x_1^p, x_2^p, x_3^p, \dots x_n^p; y^p\right)$, determine the membership values of $x_i^p (j = 1, 2, \dots N_i)$ in fuzzy set A_i^j $(j = 1, 2, \dots N_j)$, determine the membership values of output y^p in fuzzy set $B^l (l = 1, 2, \dots, N_j)$. Then, for each input variable x_i, determine the fuzzy set A_i^{j*} in which x_i^p has the maximum membership value. Similarly, the fuzzy set B^{j*} for output can also be determined.

The obtained IF-THEN Rule l is in the form of

$$\text{If } x_1 \text{ is } A_1^l \text{ and } x_2 \text{ is } A_2^1 \text{ and} \dots x_n \text{ is } A_n^l, \text{ then } y \text{ is} \tag{3}$$

The degree of rule l generated from data pair (x^p, y^p) is calculated as:

$$D(Rule) = \prod_{i=1}^{n} \mu_{A_i^j}(x_i^p) \mu_{B^j}(y^p) \tag{4}$$

2.6   Create the Fuzzy Rule Base and Construct the Fuzzy System

By accumulating the rules generated from all input-output pairs available, a desired rule base is created. This forms a look-up table. So the entire process is traditionally viewed as a table look-up scheme. Base on such rule base, a workable fuzzy system can be eventually constructed with the suitable inference engine, fuzzifier, and defuzzifier defined.

2.7   Procedure of Fuzzy Based Modeling of Yearly Change of CO_2 Due to Congestion

The nonlinear system constructed in fuzzy look-up scheme will yield out the factors which affect the change of excess CO_2 emissions due to congestion from light-duty vehicles as long as all needed demographic variables are prepared. This fuzzy table look-up scheme based modeling procedure can be summarized into the following six steps.

- Step 1: Select Sample Cities. The reasons the 96 cities are chosen in our analysis are twofold. First, these cities are located among the most populous in the United States. Second, these cities are located in the most economically vibrant areas in the United States, where population–weighted density, vehicle miles travelled, annual excess fuel consumed and excess CO_2 due to congestion change a lot these years.
- Step 2: Prepare Data Pair. The required input and output data pairs should be prepared such as the yearly change of population-weighted density, the yearly change of daily vehicle miles traveled, the yearly change of annual excess fuel consumed due to congestion, the yearly change of excess CO_2 emissions due to congestion.
- Step 3: Define Membership Functions. Membership functions for each input and output variable should be defined as required by generating input and output pairs. Small (S), Medium (M), and Large (L) will depend on the collected data and define in the fuzzy table.
- Step 4: Create Fuzzy Rule Base. Based on all available input and output paired, the fuzzy rule base will be created, each of them are similar to Eq. (3).
- Step 5: Construct Fuzzy Table. Base on the rule base from step 4, a fuzzy look-up table can be constructed.
- Step 6: Check the Modeling Error. Using a different set of input-output data pair to check the modeling error. If the errors are larger than the allowable range, necessary portion(s) of fuzzy modeling should be adjusted.

3   Data Preparation

All data used in this paper are from 2012 Annual Urban Mobility Report in Texas A&M Transportation Institute and United States Census Bureau. The report builds on previous Urban Mobility Reports with an improved methodology and expanded coverage of the nation's urban congestion problem and solutions [10]. The 2012 Urban Mobility Report builds on previous Urban Mobility Reports with an improved methodology and expanded coverage of the nation's urban congestion problem and solutions. They provide information on long-term congestion trends, the most recent congestion comparisons and a description of many congestion improvement strategies. More importantly for this paper, it has very comprehensive city-level data on variables, such as population-weighted density, vehicle miles travelled, annual excess fuel consumed, number of rush hours as well as excess CO_2 due to congestion (Million Pounds), unavailable in other databases.

4   Fuzzy Model Development

In our sample, data for 96 cities from United States are collected. The yearly change of population-weighted density (Per Square Mile) in 2000 and 2010, the yearly change of daily vehicle miles travelled (Thousand Miles) in 2000 and 2010, the yearly change of excess fuel consumed (Thousand Gallons) in 2000 and 2010, the yearly change of excess CO_2 due to congestion (Million Pounds) in 2000 and 2010 are calculated, for 96 cities out of around 20,000 cities in the United States.

4.1   Preparing Input-Output Data Pairs and Membership Function

All input variables can be indirectly obtained from the collected data as well as the output data. The membership functions for all input and output variables were selected as the typical pseudo-trapezoid membership functions with each variable classified into three fuzzy sets: Small (S), Medium (M) and Large (L). Here, 88 of 96 cities are randomly chosen. Figure 3 illustrates membership functions of three variables as input and one as output, where all input and output values are be classified into three fuzzy sets.

- The yearly change of population-weighted density between 2000 and 2010 is divided to three sets. Set S means the yearly change is less than 0, Set M means the yearly change is between 0 and 30 while Set L means the yearly change is more than 30.
- The yearly change of daily vehicle miles travelled between 2000 and 2010 is divided to three sets. Set S means the yearly change is less than 50, Set M means the yearly change is between 50 and 300 while Set L means the change is more than 300.
- The yearly change of annual excess fuel consumed between 2000 and 2010 is divided to three sets. Set S means the yearly change is less than 0, Set M means the yearly change is between 0 and 250 while Set L means the yearly change is more than 250.
- The yearly change of excess CO_2 due to congestion between 2000 and 2010 is divided to three sets. Set S means the yearly change is less than 0, Set M means the yearly change is between 0 and 20 while Set L means the yearly change is more than 20.

Then the fuzzy sets were followed to establish the rule base from the 88 cities. There are 20 rules in total in the rule base.

4.2   Fuzzy Rule Base

The 88 data pairs yielded out the IF-THEN rule base on Table 1. This would be the basis of the fuzzy model reflecting the relationships between the three input variables (IF part) and the output (THEN part) variable. The rest 8 pairs were used for validation later on.

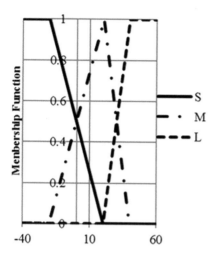

(a) Yearly Change of Population-Weighted Density

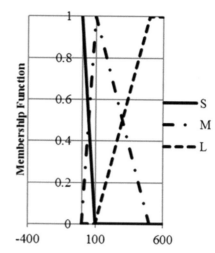

(b) Yearly Change of Daily Vehicle Miles Traveled

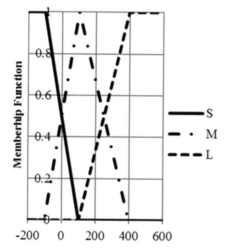

(c) Yearly Change of Excess Fuel Consumed due to Congestion

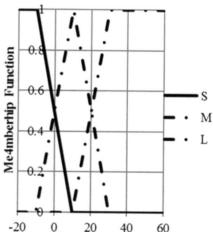

(d) Yearly Change of Excess CO_2 Emission due to Congestion

Fig. 3. Membership functions for input-output data pairs

The proposed fuzzy look-up table based method was implemented in a MS Excel spreadsheet with suitable operations. Even though these rules were based on the input-output data pairs from 88 cities in the USA, the created rule base could cover typical cities since these cities were carefully selected.

Table 1. Developed rule base to estimate change of CO_2 due to congestion

Rule	IF			THEN
	Yearly change of population-weighted density is	Yearly change of daily VMT is	Yearly change of excess fuel consumed due to congestion is	Yearly change of excess CO_2 emissions due to congestion is
1	S	S	S	M
2	S	S	M	S
3	S	M	S	S
4	S	M	M	M
5	S	M	L	M
6	S	L	S	M
7	S	L	M	M
8	S	L	L	M
9	M	S	S	S
10	M	S	M	S
11	M	S	L	S
12	M	M	S	S
13	M	M	M	M
14	M	M	L	S
15	M	L	S	M
16	M	L	M	M
17	M	L	L	M
18	L	M	L	M
19	L	M	M	M
20	L	L	M	M

4.3 Model Validation

In order to validate the constructed Fuzzy rule base and model, 5 Rule Numbers in Table 1 (Rule Number 16 MLM, Rule Number 13 MMM, Rule Number 8 SLL, Rule Number 7 SLM, and Rule Number 6 SLS), which are evaluated by the 8 cities, given the limited number of pairs in each type, are randomly chosen.

- The number of pairs for MLM is 5 in Table 1; the number of pairs for MLM is 1 in Table 2.
- The number of pairs for MMM is 5 Table 1, the number of pairs for MMM is 1 in Table 2.
- The number of pairs for SLL is 5 Table 1, the number of pairs for SLL is 2 in Table 2.
- The number of pairs for SLM is 6 Table 1, the number of pairs for SLM is 1 in Table 2.
- The number of pairs for SLS is 10 Table 1, the number of pairs for SLS is 3 in Table 2.

Table 2. A list of cities included in the model calibration

City code	City name	Yearly change of population-weighted density (Per Square Mile)		Yearly change of daily VMT (Thousand Miles)		Yearly change of excess fuel consumed due to congestion (Thousand Gallons)		Yearly change of excess CO_2 emissions due to congestion (Million Pounds)	
		Value	Set	Value	Set	Value	Set	Value	Set
1	Riverside-San Bernardino, CA	23.92	M	800.1	L	5.7	M	14.3	M
2	Bakersfield, CA	8.03	M	62.1	M	62.3	M	2.4	M
3	San Antonio, TX	−9.50	S	475.5	L	352.7	L	7.1	M
4	Oklahoma City, OK	−1.99	S	358.9	L	593.6	L	7.4	M
5	Austin, TX	−12.78	S	645.1	L	77.7	M	10.3	M
6	Las Vegas, NV	−25.46	S	816.5	L	−1312	S	7.8	M
7	Columbus, OH	−23.04	S	520	L	−728.1	S	8.3	M
8	San Diego, CA	−17.39	S	502.5	L	−87.1	S	8.5	M

First, the value of Membership Function and the value of Set in the yearly change of excess CO_2 due to congestion are used to calculate the reported average yearly change of excess CO_2 due to congestion in the Rule Base.

Second, the 8 cities in Table 2 are used and weighted-average method is used to calculate the predicted average yearly change of excess CO_2 due to congestion.

The reported yearly change of excess CO_2 emissions due to congestion is almost a little less than the predicted yearly change of excess CO_2 emissions due to congestion,

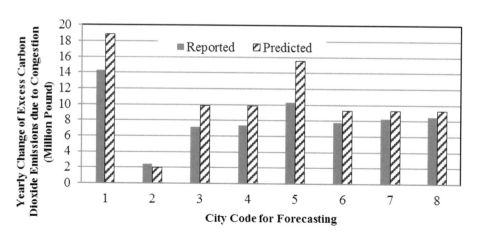

Fig. 4. Forecasting results of predicted yearly change of excess CO_2 emissions due to congestion.

which means the errors of the two is small and the developed model is practical to predict the change of excess CO_2 emissions due to congestion, can been seen. Figure 4 illustrates validation results.

5 Case Study

A case study using Austin, Texas was conducted to illustrate how the Fuzzy table look-up scheme forecasts the yearly change of excess CO_2 emissions due to congestion by 2030.

Austin, TX is the 11th most populous city in the United States of America and the fourth-most populous city in the state of Texas. Austin is also the second largest state capital in the United States. With the lower living cost and promising economic environment, more and more people and business are moving into Texas. For instance, more than 150 persons move to Austin each day, which means at least 70 more cars on the road every day (more than 25,000 additional cars a year).

The U. S. Census Bureau 2010 estimates that Austin remains one of the top destinations for migrating talent. Austin ranked first among the 50 largest U.S. metros based on net migration as a percent of total population in 2013. Because of its draw as a destination for migrating talent, Austin's population grew to nearly 1.9 million in 2013. The decade ending 2010 saw a 37% increase in population, and growth was 2.6% for the year ending July 2013. In addition, 7% of Austin residents in 2012 lived elsewhere one year earlier. That is also the largest rate among the top 50 U.S. metros. Figure 5 shows the information of population-weighted density by distance from Austin City Hall in 2000 and 2010. Figure 6 shows the information of travel time to work distribution in Austin in 2010.

5.1 Forecasting the Yearly Change CO_2 in Austin in Future Years

Three scenarios were set up to forecast the yearly change of CO_2 emissions due to congestion in Austin by 2030.

In scenario 1, the change of population-weighted density between 2005 and 2030 was assumed to be 30%; the change of daily vehicle miles traveled between 2005 and 2030 was assumed to be 130%; and the change of excess fuel consumed between 2010 and 2030 was assumed to be 88% [16].

This scenario is found in Rule 20 from the Rule Base. There are only two pairs in 88 cities. D(Rule) of this pair is 0.59041, D(IF) is 0.73499, then D(Then) is 0.8329. Through the Membership Function, the yearly change of excess CO_2 due to congestion by 2030 is 4.18%, 13.93 million pounds.

In scenario 2, the change of population-weighted density between 2005 and 2030 was assumed to be 81%; the change of daily vehicle miles traveled between 2005 and 2030 was assumed to be 10%; and the change of excess fuel consumed in 2030 was assumed to be the same as 2010 [17].

This scenario is found in Rule 19 from the Rule Base. There is only one pair in 88 cities. D(Rule) of this pair is 0.19768, D(IF) is 0.29726, then D(Then) is 0.665. Then the yearly change of excess CO_2 due to congestion by 2030 is 5.02%, 16.7 million pounds.

In scenario 3, the change of population-weighted density in 2030 was assumed to be the same as 2010; change of daily vehicle miles traveled between 2025 and 2005 was assumed to be 10%; and the change of excess fuel consumed between 2025 and 2005 was assumed to be –24% [18].

This scenario is found in Rule 19 from the Rule Base. There are only two pairs in 88 cities. D(Rule) of this pair is 0.43833, D(IF) is 0.46924, then D(Then) is 0.93413. Then the yearly change of excess CO_2 due to congestion by 2030 is 2.61%, 8.68 million pounds.

5.2 Policy Level Analysis of the Three Scenarios

As well known, different results based on different input which are defined by policy level.

For example, in one scenario, for Austin policy-makers to cap CO_2 emissions from congestion at 2.61% growth per year, or 8.68 million pounds of more from congestion than the 2010 level, the target can be achieved by keeping the population-weighted density unchanged, VMT annual growth rate capped at 2.32%, annual excess fuel consumption growth rate capped at –1.60%.

In a different scenario, for Austin policy-makers to cap CO_2 emissions from congestion at 4.18% growth per year, or 13.93 million pounds of more from congestion than the 2010 level, the target can be achieved by keeping population-weighted density annual growth rate capped at 1.20%, VMT annual growth rate capped at 3.30%, annual excess fuel consumption growth rate capped at 3.52%.

In the last scenario, for Austin policy-makers to cap CO_2 emissions from congestion at 5.02% growth per year, or 16.7 million pounds of more from congestion than the 2010 level, the target can be achieved by keeping population-weighted density annual growth rate capped at 3.24%, the daily VMT annual growth rate capped at 0.50%, annual excess fuel consumption growth rate unchanged.

Acknowledgements. The authors acknowledge that this research is supported in part by the United States Tier 1 University Transportation Center TranLIVE # DTRT12GUTC17/KLK900-SB-003, and the National Science Foundation (NSF) under grants #1137732. The opinions, findings, and conclusions or recommendations expressed in this material are those of the author (s) and do not necessarily reflect the views of the funding agencies.

References

1. U.S. Energy Information Administration: Report Number: DOE/EIA-0573 (2009)
2. Barth, M., Boriboonsomsin, K.: Traffic Congestion and Greenhouse Gases. Center of Environmental Research and Technology at the University of California-Riverside (2013)
3. Barth, M., Boriboonsomsin, K.: Real-world CO_2 impacts of traffic congestion. Transportation Research Board (2008)

4. Akar, G., Guldmann, J.-M.: Another look at VMT: determinants of vehicle use in two-vehicle households. In: Proceedings of the 91st Annual Meeting of the Transportation Research Board (2012)

5. Kopp, R.J.: Transport Policies to Reduce CO_2 Emissions from the Light-Duty Vehicle Fleet. Assessing U.S. Climate Policy Options (2012)

6. Palm, M., Brian, J.G., Wang H., McMullen, S.B.: Population density and household's transportation and housing cost trade-offs. In: 93rd Annual Meeting of the Transportation Research Board (2014)

7. Mahmood, H., Chaudary, A.R.: Population density and carbon dioxide emissions: a case study of Pakistan. Iranica J. Energy Environ. **3**(4), 354–360 (2012)

8. U.S. Census Bureau, Population Division. Metropolitan and Metropolitan Statistical Areas and Components (2009). Website: https://www.census.gov/population/metro/files/lists/2009/List1.txt

9. U.S. Energy Information Administration, Emissions of Greenhouse Gases in the U. S. (2011). http://www.eia.gov/environment/emissions/ghg_report/ghg_carbon.cfm. Accessed 8 Jan 2015

10. Texas A&M Transportation Institute, 2012 Annual Urban Mobility Report (2012). http://mobility.tamu.edu/ums/

11. Zadeh, L.A.: Fuzzy sets. Inf. Control **8**, 338–353 (1965)

12. Wang, L.-X.: A Course in Fuzzy Systems and Control. Prentice Hall PTR, Upper Saddle River (1997)

13. Qiao, F., Liu, X., Yu, L.: Arrow exit sign placement on highway using fuzzy table look-up scheme. In: Avineri, E., Koppen, M., Dahal, K., Sunitiyoso, Y., Roy, R. (eds.) Applications of Soft Computing. AISC, vol. 52, pp. 260–269. Springer Publishing (2009). ISBN: 978-3-540-88078-3

14. City of Greenville South Carolina. Driving a Modern Roundabout, South Carolina (2011). http://www.greenvillesc.gov/publicworks/Roundabouts.aspx

15. Qiao, F., Gampala, R., Yu, L.: Advanced Traffic Devices in Bicycle and Pedestrian Crossings at Freeway Interchanges. ITS World Congress, Busan, Korea (2010)

16. Tirumalachetty, S., Kockelman, K.M.: Forecasting greenhouse gas emissions from urban regions: microsimulation of land use and transport patterns in Austin, Texas. In: Transportation Research Board 89th Annual Meeting (2010)

17. Tirumalachetty, S., Kockelman, K.M., Kumar, S.: Micro-simulation models of urban regions: anticipating greenhouse gas emissions from transport and housing in Austin, Texas. In: Transportation Research Board 88th Annual Meeting (2009)

18. Musti, S., Kockelman, K.M.: Evolution of the household vehicle fleet: anticipating fleet composition, PHEV adoption and GHG emissions in Austin, Texas. Transp. Res. Part A Policy Pract. (2011)

An Experimental Case Study on Fuzzy Logic Modeling for Selection Classification of Private Mini Hydropower Plant Investments in the Very Early Investment Stages in Turkey

Burak Omer Saracoglu[✉]

Orhantepe Mahallesi, Tekel Caddesi Istanbul, Turkey
burakomersaracoglu@hotmail.com

Abstract. One of the best ways to analyze the private hydropower plant investments is to investigate them based on their installed capacities. This paper describes a proposed only one node fuzzy rule base evaluation approach (experiment and test aim) to model the selection classification of the private mini hydropower plant investments in Turkey in the very early investment stages. In these kinds of early investment stages, the data and information is generally scarcely available in a very clear, detailed, sharp and specific conditions and statuses (hence fuzzy). The total estimated electricity generation (annual), the total estimated cost (total), the change in the average surface temperature (period) are taken into account in the current experimental Mamdani's fuzzy inference based model. In this study, the mini hydropower plant investments' data were mainly gathered from the official web pages of the Republic Of Turkey Energy Market Regulatory Authority and the General Directorate of State Hydraulic Works.

Keywords: Fuzzy logic · Classification · Climate change · Experimental Fuzzy · Fuzzy inference system · Fuzzy logic toolbox · Fuzzy rule base Fuzzy rule base evaluation approach · Hydro · Hydropower · Investment Mamdani · Mini hydropower · Private investment · Private · Scilab sciFLT · Selection · Turkey

1 Introduction

When the historical electricity consumption data of Turkey is taken and analyzed, the approximately steady increase (except between the years of 2008 to 2009, 1 year) in the overall electricity consumption in Turkey is very easily observed since 2003 (data from [1]) as presented in Table 1.

After observing and understanding this reality on the electricity consumption situation in Turkey, the Turkish government and Turkish people (experts in this topic) have searched and investigated new ways to increase the amount of electricity generation in Turkey. It has been very well examined, recognized and understood that the public investments and the public investment models by the Turkish government and Turkish people (all public) can't reach the speed and acceleration of the electricity consumption

© Springer Nature Switzerland AG 2019
B. K. Ane et al. (Eds.): WSC 2014, AISC 864, pp. 69–80, 2019.
https://doi.org/10.1007/978-3-030-00612-9_7

Table 1. The electricity consumption (TWh: terawatt hour) in the Turkish Electricity System between 2003–2012 (Data: [1]) (visit [2] for unit of electric energy).

Year	2003	2004	2005	2006	2007	*2008*	*2009*	2010	2011	2012
Consumption (TWh)	141	150	161	175	190	*198*	*194*	210	230	242

in Turkey. Hence, new investment schemes and models in Turkey have been specifically studied for the Turkish electricity generation sector. One of the important and applicable recommendations in this subject has been several times made as to give the opportunity and freedom to the private investors to invest in the electricity generation sector in Turkey. After the definition and declaration of the legal scope for the private investments in the electricity generation sector in Turkey, the private investors (domestic and foreign investors) have tried to find the most appropriate and satisfactory hydropower plant investments for themselves. In this respect, the mini hydropower plant (MHPP) investments (MHPPIs) (0,1 Megawatt < Installed Capacity (P) $\leq$ 1 Megawatt), that need to be investigated according to its own characteristics, have been studied by mostly the domestic/local private investors in Turkey. A typical MHPP with its structural elements in a countryside is presented in Fig. 1 for better explanation of the MHPPs within the whole system approach. It is also very worth mentioning that these MHPPs are very effective and efficient in the stand-alone, remote area and off-grid power system and very helpful for the rural development.

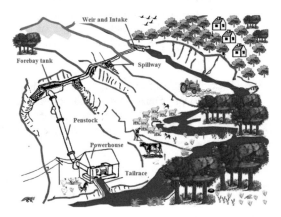

Fig. 1. Main components of MHPP scheme in a countryside (drawn, redrawn, generated and regenerated based on [3])

There are several business models, that can easily be applied in these MHPPIs. For instance, a small village and its own community can have a MHPP for their electricity generation proposes by its own in Turkey. In this study, a Virtual Private Investor (VPI) is designed, proposed and presented to analyze the available private mini hydropower plant investments (PMHPPIs) in Turkey by help of a one node Mamdani type fuzzy inference system (FIS) or Mamdani type fuzzy rule base (FRB) evaluation approach (FRBEA) (experimental).

This paper is organized in four sections. After this introduction section, the review of literature and the proposed model are presented. The experimental case study (the real world data and the evaluations of VPI) is presented. Concluding remarks and further research studies are presented in Sect. 4.

2 Literature Review

A literature review based on the predefined steps of the author's previous studies was performed on the academic publication online database and journals. Some most common terms were specifically typed and searched on these database and websites. The studies found on this literature review were tried to be presented as short as possible in a summarized way by help of Table 2. The key terms in the current study were queried as *"private mini hydropower plant investment" and "fuzzy inference system" (A), "private mini hydropower plant investment" and "Mamdani" (B), "private mini hydropower plant investment" and "fuzzy logic controller" (C), "private mini hydropower plant investment" and "Scilab" (D), "private mini hydropower plant investment" and "fuzzy logic toolbox" (E), "mini hydropower plant investment" and "fuzzy inference system" (F), "mini hydropower plant investment" and "Mamdani" (G), "mini hydropower plant investment" and "fuzzy logic controller" (H), "mini hydropower plant investment" and "Scilab" (I), "mini hydropower plant investment" and "fuzzy logic toolbox" (J), "mini hydropower plant" and "fuzzy inference system" (K), "mini hydropower plant" and "Mamdani" (L), "mini hydropower plant" and "fuzzy logic controller" (M), "mini hydropower plant" and "Scilab" (N), "mini hydropower plant" and "fuzzy logic toolbox" (O)*. It was believed that these key terms would be sufficient to present the previous studies in a detailed, but at the same time with a very narrowed set, which had very tight links between all of the documents (see Table 2).

There was only one document found on these academic publication online database and journals. This one document also was not relevant with the current study, because the key terms had been seen in the different sections of the document in the preview form. In addition to this situation, the full version of the document could not be accessible or reachable in Istanbul, Turkey due to the problems encountered by the website.

This detailed literature review showed that this subject had not most probably been studied by the researchers or any publications on this subject had not been gone to press until 01st October 2014. Under these circumstances, this study would be one of the first studies that presented a one node Mamdani type fuzzy rule base evaluation approach for the selection classification of the private mini hydropower plant investments in the very early investment stages in Turkey. Of course, very difficult questions came to the author's mind by this situation such as "what could the variables, factors, criteria be for this problem?", "how could they be fuzzified in this model?", "what could the best membership functions be in this model?". The answers to these kinds of

Table 2. Literature review summary of the current study (database and journals: ACM Digital Library-ACMDL [4], ASCE Online Research Library-ASCEOR [5], American Society of Mechanical Engineers-ASME [6], Cambridge Journals Online-CJO [7], Directory of Open Access Journals-DOAJ [8], Emerald Insight-EI [9], Google Scholar-GS [10], Journal of Industrial Engineering and Management- JIEM [11], Science Direct-SD [12], Springer-S [13], Taylor & Francis Online/Journals-TFJ [14], Wiley-Blackwell/Wiley Online Library-WB [15], World Scientific Publishing-WSP [16]).

	A	B	C	D	E	F	G	H	I	J	K	L	M	N	O
ACMDL	0	0	0	0	0	0	0	0	0	0	0	0	0	0	0
ASCEOR	0	0	0	0	0	0	0	0	0	0	0	0	0	0	0
ASME	0	0	0	0	0	0	0	0	0	0	0	0	0	0	0
CJO	0	0	0	0	0	0	0	0	0	0	0	0	0	0	0
DOAJ	0	0	0	0	0	0	0	0	0	0	0	0	0	0	0
EI	0	0	0	0	0	0	0	0	0	0	0	0	0	0	0
GS	0	0	0	0	0	0	0	0	0	0	0	0	1*	0	0
JIEM	0	0	0	0	0	0	0	0	0	0	0	0	0	0	0
SD	0	0	0	0	0	0	0	0	0	0	0	0	0	0	0
S	0	0	0	0	0	0	0	0	0	0	0	0	0	0	0
TFJ	0	0	0	0	0	0	0	0	0	0	0	0	0	0	0
WB	0	0	0	0	0	0	0	0	0	0	0	0	0	0	0
WSP	0	0	0	0	0	0	0	0	0	0	0	0	0	0	0

Note: * Irrelevant (not relevant) document. (final date of the review: 01st October 2014).

questions were tried to be given during the modeling and application of this primitive one node Mamdani type fuzzy rule base evaluation approach (experimental).

3 Proposed Experimental Mamdani's Type Model and Case Study in Turkey

The general design approach and its steps for the design process of a fuzzy rule base system was given and presented as the identification of the inputs, the identification of the outputs, the identification and decision of the membership functions, the identification and generation of the rules, the selection of the inference type and the selection of the defuzzification method [see 17, 18, 19]. The general structure of a fuzzy rule base system was also presented as shown in Fig. 2.

According to this general structure of a fuzzy rule base system and the steps for the design process of a fuzzy rule base system, the study were performed. Three factors, indices, measures and variables are included in this experimental proposed model as the total estimated electricity generation (data for average annual generation), the total estimated cost of the MHPP (data for total cost) and the change in the average surface temperature (data for period).

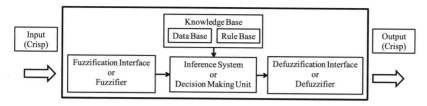

Fig. 2. Structure of a fuzzy rule base system (drawn based on [17] p. 3, [20] p. 896, [21] p. 3)

The first factor, index, measure and variable is the total estimated electricity generation. The total estimated electricity generation is usually calculated as the annual average in kWh (for kilowatt hour visit [2]) in the real life. This factor can be calculated based on the electrical installed capacity (known as the installed capacity) and the capacity factor of the MHPP. The installed capacity (in Watts) of a MHPP is calculated by the formula and Eq. 1, where η_{tr} is the efficiency of transformer, η_g is the efficiency of generator, η_t is the efficiency of turbine, ρ_w is the density of water (kg/m^3), g is the gravity (m/s^2), Q is the rated discharge (m^3/s), and H_{net} is the net head (m))

$$P = \eta tr \times \eta g \times \eta t \times \rho w \times g \times Q \times Hnet \tag{1}$$

(for extraction of this formula/equation see [22–24].

Therefore, this measure is not an independent variable, instead it is a dependent variable. The data and information for the dependent variables of this measure is not so easy to be recorded, found, taken and gathered, henceforth this index is assumed to be sufficiently usable and useful at this stage of this experimental Mamdani's type fuzzy rule base evaluation approach study. The duration and effort to find, take, gather, organize, classify and confirm these kinds of data and information is very costly (at least consider the spent manhours). During the submission of the application forms to the legal authorities, the calculated estimated average total electricity generation value is written and typed as only one value, however this is not exact in reality as it is written and typed on the application forms. Hence, it is thought in this study, that this factor can and should be modeled by help of the fuzzy theory approaches in the very early stages of the MHPP investment evaluation studies.

The second variable and factor, which is taken into account in this study, is the total estimated cost of the MHPP. This factor and index is very important, because the total amount of the capital during the application period, the procedural works period, the construction period and may be some part of the operation period is taken into account within this factor and measure. Moreover, the financial performance of the MHPP is very strictly and tightly related with this factor. This factor has a great effect on some of the investment performance indicators such as the internal rate of return, the payback period and the benefit cost ratio.

The third variable and factor, that is taken into account in the current model, is the change in the average surface temperature (the most interesting one for the author). This variable is analyzed on the basis of several years, in periods. This factor can and should be considered in these kinds of models, because of the climate change effects in our world. The climatic observations and the climate change subjected research have been conducted very seriously for almost thirty years, after the Intergovernmental Panel on Climate Change (IPCC) was established by the United Nations Environment Programme (UNEP) and the World Meteorological Organization (WMO) in 1988 [25]. The observations and the precautionary recommendations have been announced and several publications (reports, technical papers, presentations, etc.) have been presented by the IPCC since 1990 [visit 25]. Although, major actions (changes in the business models, life styles and cultures) have to be taken by the governments (governmental issues) and the people (personal issues) of the world, very few actions have been taken so far and almost no change in the life styles of the people have been observed. Henceforth, there are numerous warranty electronic mails (e-mails) have been sent and blogs have been developed considering this subject. As a consequence, the world has not been going on the right direction of being a livable home for the humankind as it should be (doing the wrong thing instead of doing the right thing), but the world had been going on the wrong direction to be a death home for the humankind (the researchers most common points on the climate change). This variable and factor is guessed to be an discriminative affect on the flow (firm flow and total flow) of the MHPPs, so that it is wise to investigate and take into account this factor as detail as possible and make models considering on this factors. The mean of annual mean surface air temperature change in the degree Celsius (°C) (visit [26] for the Celsius degrees) by the IPCC is presented as shown in Fig. 3.

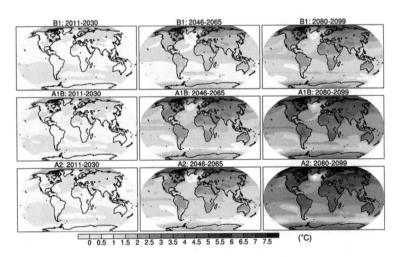

Fig. 3. The mean of annual mean surface air temperature change in °C) for three scenarios (B1, A1B and A2) for three time periods (2011–2030, 2046–2065 and 2080–2099) (Source: [27]).

After all of these inputs were identified and selected, the other steps for the other sections of the model were also followed. For instance, the linguistic fuzzy rules were defined as in the form of the most usual Mamdani's rule structure:

Rule$_i$: IF X_i is A_{i1} AND................AND X_n is A_{in} THEN Y is B_j (see [17, 18, 20, 28] where X_i: input variable/factor, A_{i1}: linguistic value of input factor, Y: output, B_j: linguistic value of output.

The mental work of this experimental proposed model was performed as the paper work, and the computational work of the current model was done by help of the Scilab Fuzzy Logic Toolbox (sciFLT) (visit [29]). The human cognition is very important in these kinds of human based/related models and studies. In the current model, the magical number 7, 7 ± 2 rule for the these cognitive issues [30, 31] (5 linguistic scales) were very seriously considered. Moreover, the Likert's research studies findings on the scales (5 Likert scales by Rensis Likert) [32] and Zadeh's research studies findings and recommendations [33] were also very much similarly tried to be adopted to the current study for the linguistic terms, hedges and verbal scales for each factor of the current model. Accordingly, the experimental model of this study was built up based on the Mamdani's approach [34] on the Zadeh's theory [35] by the following script, coded by help of the Fuzzy Toolbox 0.4.6 for the Scilab 5.4.1 (fls Editor can also be used).

Instruction	Script for 5.4.1 Scilab Editor
Note	//Create SARACOGLU WSC18 Private Mini Hydropower Plant Investments Scilab 5.4.1 SciFLT Model
Note	//Create a new fls structure.
Command	WSC18MHPPI = newfls('m','SARACOGLU WSC18 POSITION PAPER Scilab 5.4.1 SciFLT Model','asum','aprod','one','mom');
Note	//Add a new variable to the fls and return it
Note	//variable 1: total estimated electricity generation
Command	WSC18MHPPI = addvar(WSC18MHPPI,"input","total estimated electricity generation",[0 1]);
Note	//Add a new member function to the fls structure
Command	WSC18MHPPI = addmf(WSC18MHPPI,"input",1,"very low","trapmf",[0.0 0.0 0.25 0.45]); WSC18MHPPI = addmf(WSC18MHPPI,"input",1,"low","trimf",[0.25 0.45 0.55]); WSC18MHPPI = addmf(WSC18MHPPI,"input",1,"moderate","trimf",[0.45 0.55 0.65]); WSC18MHPPI = addmf(WSC18MHPPI,"input",1,"high","trimf",[0.55 0.65 0.85]); WSC18MHPPI = addmf(WSC18MHPPI,"input",1,"very high","trapmf",[0.65 0.85 1.0 1.0]);

(*continued*)

(continued)

Instruction	Script for 5.4.1 Scilab Editor
Note	//Add a new variable to the fls and return it
Command	WSC18MHPPI = addvar(WSC18MHPPI,"input","total estimated cost",[0 1]);
Note	//Add a new member function to the fls structure
Command	WSC18MHPPI = addmf(WSC18MHPPI,"input",2,"very low","trapmf",[0.0 0.0 0.2 0.4]); WSC18MHPPI = addmf(WSC18MHPPI,"input",2,"low","trimf",[0.2 0.4 0.5]); WSC18MHPPI = addmf(WSC18MHPPI,"input",2,"moderate","trimf",[0.4 0.5 0.6]); WSC18MHPPI = addmf(WSC18MHPPI,"input",2,"high","trimf",[0.5 0.6 0.8]); WSC18MHPPI = addmf(WSC18MHPPI,"input",2,"very high","trapmf",[0.6 0.8 1.0 1.0]);
Note	//Add a new variable to the fls and return it
Command	WSC18MHPPI = addvar(WSC18MHPPI,"input","change in average surface temperature",[0 1]);
Note	//Add a new member function to the fls structure
Command	WSC18MHPPI = addmf(WSC18MHPPI,"input",3,"fairly hotter","trapmf",[0.0 0.0 0.2 0.35]); WSC18MHPPI = addmf(WSC18MHPPI,"input",3,"rather hotter","trimf",[0.2 0.35 0.5]); WSC18MHPPI = addmf(WSC18MHPPI,"input",3,"hotter","trimf",[0.35 0.5 0.65]); WSC18MHPPI = addmf(WSC18MHPPI,"input",3,"very hotter","trimf",[0.5 0.65 0.8]); WSC18MHPPI = addmf(WSC18MHPPI,"input",3,"extremely hotter","trapmf",[0.65 0.8 1.0 1.0]);

Note
Command
Note
	//Add a new variable to the fls and return it
	WSC18MHPPI = addvar(WSC18MHPPI,"output","selected or unselected",[0 1]);
	//Add a new member function to the fls structure

Command	WSC18MHPPI = addmf(WSC18MHPPI,"output",1,"unselected","trapmf",[0.0 0.0 0.25 0.7]); WSC18MHPPI = addmf(WSC18MHPPI,"output",1,"selected","trapmf",[0.25 0.7 1.0 1.0]);
Note	//Plot the fls input(s) or output(s) variable(s)
Command	scf();clf();
Command	plotvar(WSC18MHPPI,"input",[1 2 3]);
Command	scf();clf();
Command	plotvar(WSC18MHPPI,"output",1);
Note	//Add 31 rules and display them in verbose format.

(continued)

(continued)

Instruction	Script for 5.4.1 Scilab Editor
Command	WSC18MHPPI = addrule(WSC18MHPPI,[1 0 0 1 1 1; 2 0 0 1 1 1; 0 5 0 1 1 1; 0 4 0 1 1 1; 0 0 5 1 1 1; 0 0 4 1 1 1; 3 3 0 1 1 1; 3 2 3 1 1 1; 3 2 2 2 1 1; 3 2 1 2 1 1; 3 1 3 1 1 1; 3 1 2 2 1 1; 3 1 1 2 1 1; 4 3 3 1 1 1; 4 3 2 2 1 1; 4 3 1 2 1 1; 4 2 3 1 1 1; 4 2 2 2 1 1; 4 2 1 2 1 1; 4 1 3 2 1 1; 4 1 2 2 1 1; 4 1 1 2 1 1; 5 3 3 2 1 1; 5 3 2 2 1 1; 5 3 1 2 1 1; 5 2 3 2 1 1; 5 2 2 2 1 1; 5 2 1 2 1 1; 5 1 3 2 1 1; 5 1 2 2 1 1; 5 1 1 2 1 1])
Note	//Show the fls rules
Command	printrule (WSC18MHPPI);
Note	//Plot the output as a function (the surface view: 3D) of the two inputs.
Command	scf();clf();
Command	plotsurf(WSC18MHPPI);

Note: There was a bug on the Scilab 5.5.0 by the 02th October 2014. Please run on the Scilab 5.4.1. until the bug shall be fixed for the Scilab 5.5.0. Please find the Scilab 5.4.1 *.fls file and the surface views of the model in the enclosed files (presentation file).

An experimental case study for the 8 PMHPPIs options (O1 to O8) in Turkey was executed and their classification as "Selected" or "Unselected" were defined. The data and information for these options were gathered from several sources (see [36–38]). The values were normalized by the maximum value due to the modeling in this study. The change in average surface temperature's values were not directly taken by any means of the software tools, but they were read from the map by the natural eye. It was imagined that the developed and sophisticated systems based on this system would get these values directly by help of the computer based systems from the forecasting studies related with this factor. The values of the other measures were also taken from the author's previous studies. The script for the evaluation of the current study was directly typed on the Scilab 5.4.1, whereas the external inputs files and outputs file were planned for the general applications. The O1, O3, O6, O7 were classified as the "Unselected" and the O2, O4, O5, O8 were classified as the "Selected". In the practical daily life, these results would tend to work on the "Selected" options in the further investment stages.

Instruction	Script
Note	//Get the SARACOGLU WSC18 Private Mini Hydropower Plant Investments Scilab 5.4.1 SciFLT Model
Command	fls = loadfls(Please enter your file path here() + "SARACOGLU WSC18 POSITION PAPER PROPOSAL 01 Rev 07");
Note	//Evaluate
Command	O1 = evalfls([0.20 0.80 0.20],fls) O2 = evalfls([0.90 0.30 0.20],fls) O3 = evalfls([0.30 0.60 0.30],fls) O4 = evalfls([0.85 0.30 0.30],fls) O5 = evalfls([0.95 0.25 0.20],fls) O6 = evalfls([0.30 0.35 0.30],fls) O7 = evalfls([0.25 0.45 0.35],fls)

(continued)

(continued)

Instruction	Script
	O8 = evalfls([0.95 0.45 0.10],fls)
Results	O1 = 0.13, O2 = 0.845, O3 = 0.13, O4 = 0.845, O5 = 0.845,
	O6 = 0.13, O7 = 0.13, O8 = 0.845

Note: There was a bug on the Scilab 5.5.0 by the 02th October 2014. Please run on the Scilab 5.4.1. until the bug shall be fixed for the Scilab 5.5.0. Please find the Scilab 5.4.1 *.fls file and the surface views of the model in the enclosed files (presentation file).

4   Conclusions and Future Work

The current study presents the possibility of using the Mamdani's type FIS for solving these kinds of problems. The future studies should focus on developing very effective and sophisticated tools based on the multi mode Mamdani's type FRB evaluation models by help of firstly the Scilab and the Scicos. In the more developed models of this study, all variables should be found, identified and analyzed very well, and the real world case studies should be finalized to help the war for the rural development (increase, goes up), the malnutrition, poverty and unemployment (decrease, go down).

Acknowledgments. The author would like to thank to Dr. Bernadetta Kwintiana Ane (conference) and Dr. Chin Luh Tan (sciFLT and Scilab 5.5.0 bug). This study would never be finalized and submitted to the conference without their consideration, guidance, and help. Please send your comments, feedbacks and criticisms to my e-mail (burakomersaracoglu@hotmail.com) in any format at any time. Your feedback will be very important and valuable for me during the development process of the models and systems for the real life applications.

References

1. Turkish Electricity Transmission Corporation: Research, Planning and Coordination Department: The Turkish Electrical Energy 5 Year Generation Capacity Projection 2013–2017 Report. Ankara, Turkey (2013)
2. Wikipedia: Wikimedia Foundation Inc. http://en.wikipedia.org/wiki/Kilowatt_hour
3. Mini Hydro Power Basics: w3.tm.tue.nl/fileadmin/tm/TDO/Indonesie/Hydro_Power.pdf
4. ACM Digital Library. http://dl.acm.org
5. ASCE Online Research Library. http://ascelibrary.org
6. American Society of Mechanical Engineers http://asmedigitalcollection.asme.org
7. Cambridge Journals Online. http://journals.cambridge.org
8. Directory of Open Access Journals. http://doaj.org
9. Emerald Insight. http://www.emeraldinsight.com
10. Google Scholar. http://scholar.google.com.tr
11. J. Ind. Eng. Manage. http://www.jiem.org/index.php/jiem
12. Science Direct. http://www.sciencedirect.com
13. Springer. http://www.springer.com/?SGWID=5-102-0-0-0
14. Taylor & Francis Online/Journals. http://www.tandfonline.com
15. Wiley-Blackwell/Wiley Online Library. http://onlinelibrary.wiley.com

16. World Scientific Publishing. http://www.worldscientific.com
17. Kumar, S., Narula, P., Ahmed, T.: Knowledge extraction from numerical data for the Mamdani type fuzzy systems: a BBO approach. In: Innovative Practices in Management and Information Technology for Excellence, Jagadhri, India (2010). http://embeddedlabcsuohio.edu/BBO/BBO_Papers/Knowledge%20Extraction.pdf
18. Noor, N.M.M., Retnowardhani, A.: An integrated framework of decision support system in crime prevention. In: DSS 2.0 – Supporting Decision Making with New Technologies, 17th Conference for IFIP WG8.3 DSS, 2–5 June 2014, Paris, France (2014). http://dss20conference.files.wordpress.com/2014/05/noor.pdf
19. Keshwani, D.R., Jones, D.D., Meyer, G.E., Brand, R.M.: Rule-based Mamdani-type fuzzy modeling of skin permeability. In: Biological Systems Engineering: Papers and Publications. Paper 80 (2008). http://digitalcommons.unl.edu/biosysengfacpub/80
20. Cordón, O.: A historical review of evolutionary learning methods for Mamdani-type fuzzy rule-based systems: designing interpretable genetic fuzzy systems. Int. J. Approximate Reasoning 52, 894–913 (2011)
21. Garg, S., Sharma, B.R., Cohen, K., Kumar, M.: A fuzzy logic based image processing method for automated fire and smoke detection. In: 51st AIAA (American Institute of Aeronautics and Astronautics) Aerospace Sciences Meeting including the New Horizons Forum and Aerospace Exposition, 07–10 January 2013, Grapevine (Dallas/FL Worth Region), Texas (2013). http://arc.aiaa.org/doi/pdf/10.2514/6.2013-879
22. Eliasson, J., Ludvigsson, G.: Load factor of hydropower plants and its importance in planning and design. In: The 11th International Seminar on Hydro Power Plants, Hydros Future in changing Markets, 15–17 November 2000, Vienna, University of Technology, Austria (2000)
23. ESHA (European Small Hydropower Association): Guide on How to Develop a Small Hydropower Plant. Belgium (2004)
24. ESHA, Brochure on Environmental Integration of Small Hydropower Plants. Belgium (2005)
25. The Intergovernmental Panel on Climate Change (IPCC) Organization. http://ipcc.ch/organization/organization.shtml
26. Wikipedia. Wikimedia Foundation Inc. http://en.wikipedia.org/wiki/Celsius
27. IPCC Fourth Assessment Report: Climate Change 2007, Climate Change 2007: Working Group I: The Physical Science Basis. http://ipcc.ch/publications_and_data/ar4/wg1/en/figure-10-8.html
28. Lermontov, A., Yokoyama, L., Lermontov, M., Machado, M.A.S.: A fuzzy water quality index for watershed quality analysis and management, environmental management in practice. In: Broniewicz, E. (ed.) ISBN: 978-953-307-358-3. InTech (2011). https://doi.org/10.5772/20316. http://www.intechopen.com/books/environmental-management-in-practice/a-fuzzy-water-quality-index-for-watershed-quality-analysis-and-management
29. sciFLT: Fuzzy Logic Toolbox for Scilab. https://atoms.scilab.org/toolboxes/sciFLT
30. Miller, G.A.: The magical number seven, plus or minus two: some limits on our capacity for processing information. Psychol. Rev. 63, 81–97 (1956)
31. Shiffrin, R.M., Nosofsky, R.M.: Seven plus or minus two: a commentary on capacity limitations. Psychol. Rev. 101(2), 357–361 (1994)
32. Likert, R.: A technique for the measurement of attitudes. Arch. Psychol. No. 140, New York, USA (1932)
33. Zadeh, L.A.: A fuzzy-set-theoretic interpretation of linguistic hedges. J. Cybern. 2(3), 4–34 (1972)
34. Mamdani, E.H.: Application of fuzzy algorithms for control of simple dynamic plant. Proc. Inst. Electr. Eng. 121(12), 1585–1588 (1974)

35. Zadeh, L.A.: Fuzzy sets. Inf. Control **8**, 338–353 (1965)
36. General Directorate of State Hydraulic Works. http://www.dsi.gov.tr
37. Republic of Turkey Energy Market Regulatory Authority. http://www.emra.gov.tr/, http://www.epdk.gov.tr
38. Turkish Electricity Transmission Corporation. http://www.teias.gov.tr

An Experimental Study on Fuzzy Expert System: Proposal for Financial Suitability Evaluation of Commercial and Participation Banks in Power Plant Projects in Turkey

Burak Omer Saracoglu[✉]

Orhantepe Mahallesi, Tekel Caddesi, Istanbul, Turkey
burakomersaracoglu@hotmail.com

Abstract. One of the very important and critical activities of the commercial & participation banks in Turkey is the real sectors' project financing. The authorized Turkish banks, that are allowed to accept deposits and loan projects, should play a key role in the Turkish economy. The Turkish people should expect from them to manage this process in a very appropriate, transparent and trustworthy way by also considering their social responsibilities (not in favor of only net profit maximization objective). The most agreeable, applicable, beneficial, suitable, and useful decisions on "what to finance", "when to finance", "how to finance", "in what terms to finance", and "how long to finance" will make the whole system works, otherwise the country fails and collapses. This paper proposes an experimental fuzzy expert system to the commercial & participation banks for their power plant projects financing (loan) suitability evaluation studies in Turkey.

Keywords: Fuzzy logic · Bank · Climate change · Commercial bank
Credit · Decision making system · Electricity · Energy · Experimental
FuzzME · FuzzME software tool · Fuzzy expert system
Fuzzy inference system · Fuzzy models of Multiple-Criteria evaluation
Fuzzy rule base system · Fuzzy rule base evaluation system · Investment
Loan · Mamdani · Participation bank · Power · Power plant · Private investment
Private · Project financing · Sugeno · Turkey

1 Introduction

Nowadays, Turkey (Turkish economy) is placed in the most fragile economies, due to some of its demonstrative data by its own frightening indicators such as the gross domestic/national savings (% of GDP: gross domestic product) (visit [1] for the terminology), the external debt stocks (% of GNI: gross national income) (visit [2]), the total amount of credits/loans and the income share held by lowest 20% [3–7]. The historical data onto these indicators is presented in Fig. 1 for the better perspective and understanding of the actual condition. The definitions, descriptions and explanations of these measures can be accessible from the cited websites by the readers [visit 8–11]. It should be emphasized that the gross domestic/national savings measure in Turkey was the

© Springer Nature Switzerland AG 2019
B. K. Ane et al. (Eds.): WSC 2014, AISC 864, pp. 81–91, 2019.
https://doi.org/10.1007/978-3-030-00612-9_8

lowest one amongst the developing countries in February 2014 [12–15]. This indicator had been getting worse and worse for years and years and concluded with this situation as shown in Fig. 1.

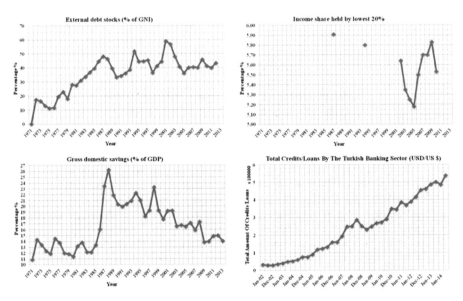

Fig. 1. The indicators of the Turkish Economy (Data: external debt stocks [8] (top left), income share held by lowest 20% [9] (top right), gross domestic/national savings [10] (bottom left), total amount of credits/loans [11] (bottom right)) (final date of the review: 01st October 2014).

On one hand the Turkish real sectors' investments' requirements have been observed, such as the electricity generation sector for urgently balancing the electricity demand of the Turkish people, on the other hand the sufficient amount of the capital to actualize for these real sectors' investments can't be found by its own resources (gross domestic/national savings). The country lives on the edge of this reality (without attracting adequate and enough foreign direct investment). Hence, the country's collapse risk at once by a very small step should always be on the minds. The Turkish banks should be very careful and take the lead (e.g. consistent, recommending the right, not directing to the wrong, morality, honest, open, and free) in these kinds of periods. They should be very aware of and take into account the bank accounts owners' capital, their own capitals and the percentages of these two important resources. On this respect, they should find the most relevant ways for themselves to find the answers of "what to finance", "when to finance", "how to finance", "in what terms to finance", and "how long to finance". The research studies in this subject will most probably very helpful to understand the problems and their corresponding answers in the near future.

In this study, a so called Virtual Commercial and Participation Bank (VCPB) in Turkey (without considering the differences of the commercial banking system and the participation or the Islamic banking/financing system) is proposed to investigate a few of the available power plant projects (PPPs) in Turkey to be financed. The "virtual" term is specifically chosen by the author to represent the readers that the author of this study models a fictional bank by himself and makes all the evaluations by himself. Only the data of the power plants are taken from the real life. In the theoretical approach, the models of the commercial banks and the participation banks are very different from each other for the real practical life. Akbank, Anadolubank, Fibabanka, Sekerbank, Tekstil Bankasi, Turkish Bank, Turk Ekonomi Bankasi, Turkiye Garanti Bankasi, Turkiye Is Bankasi, and Yapi ve Kredi Bankasi are grouped under the private commercial banks in Turkey [16]. Ziraat Bankasi, Turkiye Halk Bankasi, and Turkiye Vakiflar Bankasi are grouped under the public commercial banks in Turkey [16]. Albaraka Turk Katilim Bankasi, Asya Katilim Bankasi, Kuveyt Turk Katilim Bankasi, and Turkiye Finans Katilim Bankasi are grouped under the participation banks [16] in Turkey. The title of the current study does not mean and indicate, that the current proposed model and its advanced developed and adopted ones in the future can't be applied and executed by the other groups such as the public investment banks, the private investment banks, and the foreign investment banks; instead the more sophisticated and adopted ones can be developed by themselves [16].

A one node Mamdani type fuzzy inference system (FIS) or Mamdani type fuzzy rule base evaluation approach or fuzzy expert system is experimentally proposed for the power plant projects financing (loan) suitability evaluations in Turkey in an experimental study point of view.

This paper has four sections. The following section is for the review of literature. Section 3 presents the experimentally proposed Mamdani type fuzzy rule base evaluation approach and the virtual bank case study (the real world data and the VCPB) and the concluding remarks and further research studies are presented in Sect. 4.

2 Literature Review

The literature review was completed by help of some key terms or phrases. The key words and phrases were searched on some scientific online databases and journal websites. The queries of the current literature review were *"banking" and "fuzzy inference system" (A), "banking" and "Mamdani" (B), "banking" and "fuzzy rule base" (C), "banking" and "FuzzME" (D), "financing" and "fuzzy inference system" (E), "financing" and "Mamdani" (F), "financing" and "fuzzy rule base" (G), "financing" and "FuzzME" (H), "loan" and "fuzzy inference system" (I), "loan" and "Mamdani" (J), "loan" and "fuzzy rule base" (K), "loan" and "FuzzME" (L)*. The scientific online database and journal websites preferred in the current study were shown in Table 1 and the links were given in the references [17–23].

The total number of the documents found on the scientific online database and journal websites were 9134. Only 181 documents amongst these 9134 documents were relevant in a general manner with the current study according to the author's point of view. Most of these relevant documents were the duplications or triplications etc. of the original document on the different scientific online database and journal websites. The irrelevant documents were eliminated and shown in Table 1. The most relevant and important studies in a general manner were briefly tried to be explained in this section.

Table 1. Literature review summary of the current study: capital letters: total number of found documents, small letters: relevant studies with this study (databases and journals: ACM Digital Library-AC [17], Cambridge Journals Online-CJ [18], Directory of Open Access Journals-DO [19], Emerald Insight-EI [20], Google Scholar-GS [21], Science Direct-SD [22], Springer-S [23]); final date of the review: 03rd October 2014.

	AC	CJ	DO	EI	GS	SD	S		AC	CJ	DO	EI	GS	SD	S
A	67	0	1	4	617	94	1	G	3	0	0	0	66	7	8
a	4	0	0	0	15	4	0	g	0	0	0	0	2	0	0
B	46	100	1	4	1.710	62	0	H	0	0	0	0	0	0	0
b	2	0	0	0	17	2	0	h	0	0	0	0	0	0	0
C	33	0	0	2	217	40	10	I	43	0	0	3	402	88	0
c	0	0	0	0	18	3	0	i	5	0	0	1	27	4	0
D	0	0	0	0	2	0	0	J	26	105	0	2	1.830	88	0
d	0	0	0	0	1	0	0	j	4	0	0	0	12	4	0
E	19	0	0	2	304	56	1	K	28	0	0	0	237	34	1
e	1	0	0	0	20	3	0	k	2	0	0	0	9	2	0
F	8	113	0	3	2.590	56	0	L	0	0	0	0	0	0	0
f	0	0	0	0	19	0	0	l	0	0	0	0	0	0	0

Kumar et al. focused on the selection of loan applicants by the financial organizations and banks and developed a risk analysis based on the fuzzy inference system [24]. Oreski and Oreski presented a hybrid genetic algorithm with neural networks (HGA-NN) for the credit risk assessment [25]. Tsai et al. presented an application of a consumer loan default predicting model based on the data envelopment analysis (DEA) and the neural network (NN) for a financial institution in Taiwan [26]. Odeh et al. applied the Fuzzy Simplex Generic Algorithm for the loan decisions [27]. Eletter and Yaseen used the artificial neural network (ANN) to evaluate the loan applications in the Jordanian commercial banks [28]. Lahsasna developed a multi phase intelligent credit scoring model by help of a fuzzy rule base and a multi-objective genetic algorithm [29]. Sreekantha and Kulkarni designed a methodology for the credit risk evaluation of the enterprises for banks based on the fuzzy logic and neural network techniques [30]. Pourdarab et al. adopted the Kolmogorove-Smirnov test, the Decision Making Trial and Evaluation Laboratory (DEMATEL) method and the Mamdani's Fuzzy Expert System (inputs: current ratio, debit ratio, return on sales, average collection period, quick ratio; output: credit risk degree of customer) for the credit risk assessment of the bank customers in an Iranian Bank [31]. Malhotra and Malhotra applied the artificial neuro-fuzzy inference systems (ANFIS) and the multiple discriminant analysis models for the evaluation of the consumer loan applications [32]. Lahsasna et al. surveyed the studies used the soft computing methods for the credit scoring models [33]. Jiao et al. modeled a fuzzy adaptive network (FAN) for the credit rating of the small financial enterprises [34].

This widened literature review exposed that the researchers had been interested in the applications of the fuzzy rule base systems and the fuzzy inference systems in the financial industry and banks. However, developing a fuzzy expert system for the commercial & participation banks in the power plant projects financing (loan) suitability evaluations in Turkey and its broaden topic as developing a fuzzy rule base system for the financial institutions in the power plant projects financing (loan) suitability evaluations had not most probably been come to the researchers' and scientists' interests until 03th October 2014. This literature review presented that this study is most probably the first study in its subject. It is believed that this study will be a good foundation by its one node Mamdani type fuzzy rule base evaluation system on this scope for the future research studies.

3   Virtual Bank Case Study: The Experimentally Proposed Fuzzy Rule Base System

It is hoped that the whole development process completed sophisticated fuzzy rule base systems shall take the attention of the Turkish commercial & participation banks to deal with the ambiguity, doubtfulness, imprecision, unsharpness, unclearness, and uncertainty during their power plant projects financing suitability evaluations in Turkey [see for the descriptive terminology 35–38]. The current study is founded on the fuzzy logic and its principles, the Mamdani's fuzzy inference system and its principles, and the FuzzME software package and its capabilities. Henceforth the previous studies by Lotfi A. Zadeh, by Ebrahim H. (Abe) Mamdani and by Jana Talasova, Pavel Holecek and their colleagues should be reviewed (see [39–45].

There are three variables and factors taken into account in the current study. The preferred measures and variables are the levelized cost of the power plant type by the electricity generation technology (LCOE) (quantitative criterion), the capacity factor of the power plant (quantitative), and the loan to cost ratio (LTC) (quantitative).

The first variable in this study is the LCOE (Variable 1: V_1). The LCOE is generally defined as "*levelized cost of electricity (LCOE) is often cited as a convenient summary measure of the overall competiveness of different generating technologies. It represents the per-kilowatt hour cost (in real dollars) of building and operating a generating plant over an assumed financial life and duty cycle*" [46] (also [47–49]) and formulated as

$$LCOE = \frac{Total\,Life\,Cycle\,Cost\,(US\,Dollar,\ Euro,\ Renminbi:Yuan)}{Total\,Lifetime\,Energy\,Production\,(GWh/MWh/kWh)} \quad (1)$$

$$LCOE = \frac{Initial\,Investment - Depreciation\,Tax\,Shield + Annual\,Cost - Salvage\,Value}{Total\,Lifetime\,Energy\,Production}$$

$$(2)$$

(see extraction and details of this equation in [49–51]).

The second variable in the current study is the capacity factor (V_2), that is generally defined as "the ratio of the net electricity generated, for the time considered, to the energy, that could have been generated at continuous full-power operation during the same period" [52, 53] and formulated as

$$Net\ Capacity\ Factor\ (Power\ Plant) = \frac{Total\ Annual\ Energy\ Production\ (GWh/MWh/kWhperyear)}{(365\ days) \times (24\ h/day) \times Installed\ Capacity\ (GW/MW/kW)}$$

(3)

(see extraction and details of this equation in [53, 54]).

The third variable and metric in this study is the total percentage of the loan requested for the power plant (LTC) (V_3) and formulated as

$$LTC = \frac{Loan\ Used\ To\ Finance\ The\ Project}{Cost\ To\ Build\ The\ Project}$$

(4)

(see extraction and details of this equation in [55, 56]).

The typical units of the LCOE (V_1) is the US cents/kWh or MWh, the euro/kWh or MWh, and both the net capacity factor (V_2) and the LTC (V_3) are unitless (visit [57, 58] for learning the units). During the modeling of the linguistic terms and verbal scales of the factors, the research studies and findings of Rensis Likert (1903–1981), Lotfali Askar Zadeh (1921–alive), George Armitage Miller (1920–2012), Richard M. Shiffrin (1968–alive) and Robert M. Nosofsky (alive) and finally Bernd Rohrmann (alive) was used and adopted in this study (see Table 2). The Mamdani's fuzzy rule based system

Table 2. The representative normalized membership functions on the FuzzME Software (open presentation and FuzzME model files).

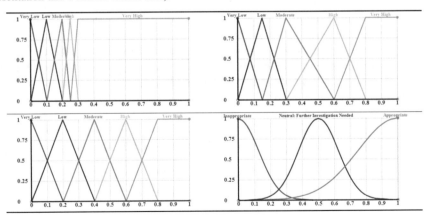

Input: LCOE V_1: Decreasing Scale lower values are better (up left), Capacity Factor V_2: Increasing Scale higher values are better (up right), LTC V_3: Decreasing Scale lower values are better (down left).
Output: Increasing Scale higher values are better (down right).
Note: Increasing Scale: values in the real life are normalized with the maximum value,
Decreasing Scale: first, values in the real life are normalized with the maximum value; second, normalized values are calculated and changed as x to (1−x) or directly decreasing scale selected on the FuzzME.

(FRBS) was preferred to be adopted in this study, because of its main agreed upon capability to model human judgments better than other FRBSs (see [59, 60]). The data and information set was normalized to be able to use the FuzzME. The current model was finalize with 50 defined rules in the standard form of the Mamdani's FIS (Fig. 2). The experimental application was performed with a few randomly selected power plants in Turkey (data from [61] and [51, 53]).

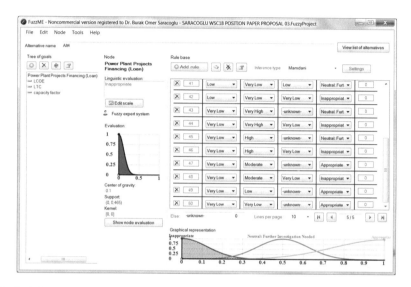

Fig. 2. The Mamdani's fuzzy rule base on FuzzME (open presentation & FuzzME files).

The LCOE (V_1) was modeled and evaluated based on the data and information of the Open Energy Information by help of the fuzzy numbers for the criterion value according the type of the power plant on the FuzzME (visit [63]). The net capacity factor (V_2) was calculated based on the data and information of the Turkish Electricity Transmission Corporation and modeled and evaluated as the positive real numbers (IR^+). The LTC (V_3) was appointed randomly between 0% and 80% due to the capital requirements of the EMRA (Republic of Turkey Energy Market Regulatory Authority) for the power plant investments (20% investors' capital) with fuzzy numbers. The power plant alternatives were in the natural gas, the hydropower, the wind, the biogas, and the coal segments. The evaluations for these 6 alternatives were presented in Table 3. The final results were gathered only for two groups as "Inappropriate to Neural: Further Investigation Needed" for Alt4 and Alt6 and "Inappropriate" for Alt1, Alt5, Alt2 and Alt3 (Alt: alternative). These results indicated that alternatives 4 and 6 had to be investigated in detail, but the others had to be not investigated and not financed.

Table 3. The evaluations of alternatives (criterion value) on the FuzzME Software (open presentation and FuzzME model files)

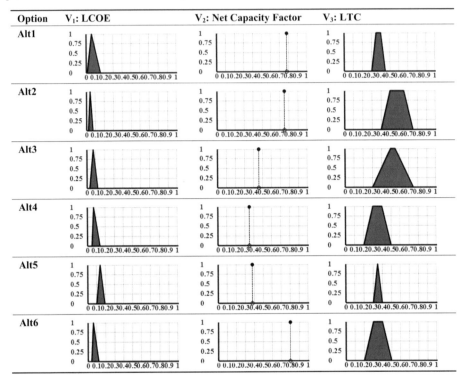

4 Conclusions and Future Work

It is hoped that this study can present the possibility of using the Mamdani type fuzzy inference system in this very important and crucial subject. If the sophisticated multi mode Mamdani type fuzzy rule base evaluation models on the FuzzME can be built up in the short to mid term, the answers of the questions can be given with ease in a proper and transparent way, which will of course help the country to be a livable home. The future studies should cover the subjects of identifying and defining the variables, the fuzzification of these defined variables and the definition of the rules to develop the systems (a unique system for a bank) for the private commercial banks, the public commercial banks and the participation banks (including others also) to help them in their evaluations in the real world applications.

Acknowledgments. The author would like to thank to Dr. Bernadetta Kwintiana Ane (conference) and Dr. Pavel Holeček (FuzzME). This study shall never be finalized and submitted to the conference without their consideration, guidance, and help. Please send your comments, feedbacks and criticisms to my e-mail (burakomersaracoglu@hotmail.com) in any format at any time. Your feedback will be very important and valuable for me during the development process of the models and systems for the real life applications.

References

1. Wikipedia: Wikimedia Foundation Inc., gross domestic product (GDP). http://en.wikipedia.org/wiki/Gross_domestic_product
2. Wikipedia: Wikimedia Foundation Inc., Gross National Income (GNI). http://en.wikipediaorg/wiki/Gross_national_income
3. Trading Economics: Turkey Economic Indicators. http://www.tradingeconomics.com/turkey/indicators
4. Fitch Ratings Inc.: Bank Borrowing Drives Rise in Turkey's External Debt. https://www.fitchratings.com/gws/en/fitchwire/fitchwirearticle/Bank-Borrowing-Drives?pr_id=863194
5. Thomson Reuters, Fitch: Bank Borrowing Drives Rise in Turkey's External Debt. http://www.reuters.com/article/2014/09/03/fitch-bank-borrowing-drives-rise-in-turk-idUSFit72652120140903
6. Anadolu Agency: Turkey's private sector foreign debt rises $8.9 billion. http://www.aacom.tr/en/economy/376118–turkeys-private-sector-foreign-debt-rises-8-9-billion
7. Forbes.com LLC: Why Bubble Warnings Aren't Immediate 'Sell' Signals. http://www.forbes.com/sites/jessecolombo/2014/08/31/why-bubble-warnings-arentimmediate-sell-signals
8. The World Bank Group: The external debt stocks (% of GNI). http://data.worldbank.org/indicator/DT.DOD.DECT.GN.ZS/countries/TR?display=graph
9. The World Bank Group: The income share held by lowest 20%. http://data.worldbank.org/indicator/SI.DST.FRST.20/countries/TR?display=graph
10. The World Bank Group: The gross domestic savings (% of GDP). http://data.worldbank.org/indicator/NY.GDS.TOTL.ZS/countries/TR?display=graph
11. The Banking Regulation and Supervision Agency (BBDK): Weekly Reports (e-bulletin/Interactive Weekly Bulletin). http://ebulten.bddk.org.tr/haftalikbulten/BultenYeni.aspx
12. Today's Zaman: Reports: Turkey has worst savings rate of emerging economies. http://www.todayszaman.com/news-340565-reports-turkey-has-worst-savings-rate-of-emerging-economies.html
13. Rabobank: The economic research department. https://economics.rabobank.com/publications/2014/march/country-report-turkey
14. Bloomberg: Basci sees deficit salvation with savers blessing: Turkey. http://www.bloomberg.com/news/2014-09-11/basci-sees-deficit-salvation-with-savers-blessingturkey.html
15. Balkans Business News: Turkey's progress towards economic rebalancing may become more challenging for the remainder of 2014. http://www.balkans.com/open-news.php?uniquenumber=196916
16. The Banking Regulation and Supervision Agency (BDDK): Banks http://www.bddk.gov.tr/WebSitesi/english/Institutions/Banks/Banks.aspx
17. ACM Digital Library. http://dl.acm.org
18. Cambridge Journals Online: http://journals.cambridge.org
19. Directory of Open Access Journals. http://doaj.org
20. Emerald Insight. http://www.emeraldinsight.com
21. Google Scholar. http://scholar.google.com.tr

22. Science Direct. http://www.sciencedirect.com
23. Springer. http://www.springer.com/?SGWID=5-102-0-0-0
24. Kumar, S., Bhatia, N., Kapoor, N.: Fuzzy logic based decision support system for loan risk assessment. In: ACAI 2011 Proceedings of the International Conference on Advances in Computing and Artificial Intelligence, pp. 179–182 (2011). https://doi.org/10.1145/2007052. 2007089
25. Oreski, S., Oreski, G.: Genetic algorithm-based heuristic for feature selection in credit risk assessment. Expert Syst. Appl. Part 2 **41**(4), 2052–2064 (2014). https://doi.org/10.1016/j. eswa.2013.09.004
26. Tsai, M.C., Lin, S.P., Cheng, C.C., Lin, Y.P.: The consumer loan default predicting model - an application of DEA-DA and neural network. Expert Syst. Appl. **36**(9), 11682–11690 (2009). https://doi.org/10.1016/j.eswa.2009.03.009
27. Odeh, O., Koduru, P., Featherstone, A., Das, A., Welch, S.M.: A multi-objective approach for the prediction of loan defaults. Expert Syst. Appl. **38**(7), 8850–8857 (2011). https://doi. org/10.1016/j.eswa.2011.01.096
28. Eletter, S.F., Yaseen, S.G.: Applying neural networks for loan decisions in the Jordanian
29. Commercial Banking System: IJCSNS Int. J. Comput. Sci. Netw. Secur. **10**(1), 209–214 (2010). http://paper.ijcsns.org/07_book/201001/20100128.pdf
30. Lahsasna, A.: Intelligent credit scoring model using soft computing approach. In: Proceedings of the International Conference on Computer and Communication Engineering 2008 (ICCCE08), pp. 396–402 (2008). http://dx.doi.org/10.1109/ICCCE.2008.4580635
31. Sreekantha, D.K., Kulkarni, R.V.: Expert system design for credit risk evaluation using neuro-fuzzy logic. Expert Syst. **29**, 56–69 (2012). https://doi.org/10.1111/j.1468-0394.2010. 00562
32. Pourdarab, S., Nadali, A., Nosratabadi, H.E.: A hybrid method for credit risk assessment of bank customers. Int. J. Trade Econ. Financ. **2**(2), 125–130 (2011). http://www.ijtef.org/ papers/90-F00091.pdf
33. Malhotra, R., Malhotra, D.K.: Differentiating between good credits and bad credits using neuro-fuzzy systems. Eur. J. Oper. Res. **136**(1), 190–211 (2002)
34. Lahsasna, A., Ainon, R.N., Wah, T.Y.: Credit scoring models using soft computing methods: a survey. Int. Arab J. Inf. Technol. **7**(2), 115–123 (2010)
35. Jiao, Y., Syau, Y.R., Lee, E.S.: Modelling credit rating by fuzzy adaptive network. Math. Comput. Model. **45**(5–6), 717–731 (2007). https://doi.org/10.1016/j.mcm.2005.11.016
36. Wierman, M.J.: An Introduction to the Mathematics of Uncertainty including Set Theory, Logic, Probability, Fuzzy Sets, Rough Sets, and Evidence Theory. Center for the Mathematics of Uncertainty, Creighton University College of Arts and Sciences (2010)
37. Klir, G.J., Yuan, B.: Fuzzy Sets and Fuzzy Logic Theory and Applications. Prentice Hall, Upper Saddle River (1995)
38. Kasabov, N.K.: Foundations of Neural Networks, Fuzzy Systems, and Knowledge Engineering. MIT Press (1998)
39. Sowa, J.F.: What is the source of fuzziness? Stud. Fuzziness Soft Comput. **299**, 645–652 (2013)
40. Zadeh, L.A.: Fuzzy sets. Inf. Control **8**, 338–353 (1965)
41. Zadeh, L.A.: Communication fuzzy algorithms. Inf. Control **12**, 94–102 (1968)
42. Zadeh, L.A.: Toward a perception-based theory of probabilistic reasoning with imprecise probabilities. J. Stat. Planning Infer. **105**, 233–264 (2002)
43. Zadeh, L.A.: Generalized theory of uncertainty (GTU) principal concepts and ideas. Comput. Stat. Data Anal. **51**, 15–46 (2006)
44. Mamdani, E.H.: Application of fuzzy algorithms for control of simple dynamic plant. Proc. Inst. Electr. Eng. **121**(12), 1585–1588 (1974)

45. Talasova, J., Holecek, P.: Multiple-Criteria Fuzzy Evaluation: the FuzzME Software Package. IFSA/EUSFLAT Conf. **2009**, 681–686 (2009)
46. Fuzzy MCDM: Research team of the Department of Mathematical Analysis and Applications of Mathematics, Faculty of Science, Palacký University Olomouc, Olomouc, Czech Republic, Selected publications. http://fuzzymcdm.upol.cz/index.php/publications
47. U.S. Energy Information Administration, http://www.eia.gov/forecasts/aeo/electricity _generation.cfm
48. U.S. Department of Energy: Office of Energy Efficiency and Renewable Energy, National Renewable Energy Laboratory (NREL). http://www.nrel.gov/analysis/tech_lcoe.html
49. Campbell, M., Aschenbrenner, P., Blunden, J., Smeloff, E., Wright, S.: The drivers of the levelized cost of electricity for utility-scale photovoltaics. SunPower Corp. (2008)
50. Borenstein, S.: The Private And Public Economics Of Renewable Electricity Generation, National Bureau Of Economic Research, Cambridge. http://www.nber.org/ papers/w17695 (2011)
51. Vasudev, A.: The Levelized Cost of Electricity. http://large.stanford.edu/courses/2010/ ph240/vasudev1
52. Wikimedia Foundation Inc.: Wikipedia, Cost of electricity by source. http://en.wikipedia. org/wiki/Cost_of_electricity_by_source
53. U.S. Nuclear Regulatory Commission: NRC Library. http://www.nrc.gov/reading-rm/basic-ref/glossary/capacity-factor-net.html
54. Wikimedia Foundation Inc.: Wikipedia, Capacity factor. http://en.wikipedia.org/wiki/ Capacity_factor
55. Nyboer, J., Lutes, K.: A review of renewable energy in Canada, 2009. In: Natural Resources Canada and Environment Canada, Canadian Industrial Energy End-use Data and Analysis Centre Simon Fraser University, Burnaby, BC (2011)
56. Investopedia US: http://www.investopedia.com/terms/l/loan-to-cost-ratio-ltc.asp
57. Fundrise LLC: https://fundrise.com/education/glossary/loan-to-cost-ratio-ltc
58. Wikipedia: Wikimedia Foundation Inc., Kilowatt hour. http://en.wikipedia.org/wiki/ Kilowatt_hour
59. Wikipedia: Wikimedia Foundation Inc. http://en.wikipedia.org/wiki/Currency
60. Cordon, O.: A historical review of evolutionary learning methods for Mamdani-type fuzzy rule-based systems: designing interpretable genetic fuzzy systems. Int. J. Approximate Reasoning **52**, 894–913 (2011)
61. Alcala, R., Casillas, J., Cordon, O., Herrera, F., Zwir, S.J.I.: Techniques for learning and tuning fuzzy rule-based systems for linguistic modeling and their application
62. Turkish Electricity Transmission Corporation: Research Planning and Coordination Department, Turkish Electrical Energy 10-Year Generation Capacity Projection 2009–2018 (2009)
63. Open Energy Information (OpenEI): Transparent Cost Database, LCOE. http://en.openei. org/apps/TCDB

Evolutionary Methods

Genetic Algorithms for the Non Emergency Patients Transport Service

José A. Oliveira$^{(\boxtimes)}$, João Ferreira, Manuel Figueiredo, Luis S. Dias, and Guilherme A. B. Pereira

ALGORITMI Research Centre, University of Minho, Braga, Portugal
jose.oliveira@algoritmi.uminho.pt,
joao.aoferreira@gmail.com,
{mcf,lsd,gui}@dps.uminho.pt,
http://pessoais.dps.uminho.pt

Abstract. This study presents three genetic algorithms developed to solve a routing problem related with a transportation service for nonemergency patients in Portugal. A model based on an extension of the Team Orienteering Problem was developed to carry out several legal constraints. The results of computational experiments made to validate the adopted methodology, using both public TOP instances and real data, are presented.

Keywords: Genetic Algorithm · JCell framework
Team Orienteering Problem · Non-emergency patients transport services

1 Introduction

This paper presents three genetic algorithms to define routes related with the non-emergency patients transport service provided in Portugal. The transport service is free for disadvantaged people and is paid by the National Health Service. The transport service must be done with the necessary quality and safety, which impose several constraints to be fulfilled when the routes are planned. Ideally, the patient transport service would carry one patient at a time on each route. This individualized transport would be faster and more comfortable service for the patient. However, this would greatly increase the necessary resources, since it would be necessary to use more vehicles and recruit more crew members. The costs increase disproportionately, whether direct costs or environmental costs. On the other hand, considering that the need for health care is increasing and the number of disadvantaged people also rises, the budget constraint is of huge importance to the national authorities. Rationalizing the costs of the patient transport service assumes a critical role for the National Health Service. To rationalize these costs, the transport can be made by grouping more patients on the same route, using vehicles with greater capacity and traveling longer distances. The drawback to the patient is to travel more distance than is strictly necessary to take him to his destination, increasing the travel time and waiting for other patients. It is necessary to find a balance between these two cases, where the individual transport should be avoided, and overly large routes should also be avoided. A team of experts consisting of doctors, nurses, managers, ambulance drivers conducted a study on the

B. K. Ane et al. (Eds.): WSC 2014, AISC 864, pp. 95–106, 2019.
https://doi.org/10.1007/978-3-030-00612-9_9

subject of transport of non-urgent patients and defined the limits for the maximum length of the routes, and other rules. Taking into account this study new legislation concerning this service was published in 2012 as part of efforts to reduce the cost of health care [1, 2]. The new legislation drew our attention to the subject and we decided to model the real problem as an optimization problem, in which rationalizing the costs of the patients transport service assumes a critical role. This situation drives to a large-scale real-world optimization problem, and to deal with this task the use of heuristics is mandatory. Our experience with genetic algorithms encouraged us to develop a new approach to this optimization problem to produce efficient solutions.

This type of service can be seen as an ambulance transport sharing problem, and similar strategies developed in the context of carpooling (or bus pooling or taxi pooling) could be implemented. This paper is structured in the following way: Sect. 2 describes the TOP-based model and justifies the adopted strategy. Section 3 discusses the selected methodology and describes the "GAs" implementation. The obtained computational results are presented in Sect. 4. Some conclusions are summarized in Sect. 5.

2   Modeling the Real Problem

In the Western world in general, and in Portugal in particular, demand for health services is increasing, because in these societies the aging population is large, and the elderly are increasingly isolated and alone. The state of health of patients does not allow them to move by its own means to obtain the health care, which worsens their health conditions [3]. The isolation in which they live and the dispersion of their homes are geographic factors that make access to the treatments difficult. So, it is important that the state provide free transport, while ensuring that the cost of the transport service is controlled.

The non-emergency patient transport is a free service in Portugal to the economically disadvantaged people and is paid for by the National Health Service. This service in 2012 had a very high cost and so it was necessary to implement measures to reduce this cost. Our approach to the problem assumes that all patients are transported to their destination, and we establish individual transportation as the initial solution. Then we will select patients who can travel together. Patients who are not selected at this stage, will be transported individually in the final solution. The stage of the selection of patients to be transported together is similar to the choice of vertices in a graph for the team orienteering problem (TOP) [4]. In the TOP not all vertices are visited. In this real-world problem we assume that not all patients are chosen to travel in a shared route. We model this problem as a routing problem with several particularities based in the Team Orienteering Problem. In this problem, considering the available set of patients to be transported, we select some patients to be transported by the planned routes, leaving the other patients without being transported in these routes. The selection of transported patients is made taking into account the costs saved. For the non-visited patients, it is necessary to provide them with individual transportation.

We obtain a savings by grouping two patients as follows. Figure 1 illustrates two cases to collect two patients. In the first case two vehicles are used, each collecting one patient, and in the second case one vehicle is used to collect both patients, but patient A is collected in first place, before patient B. Patient A is called "first" patient. According to Portuguese law, he must be the more distant patient.

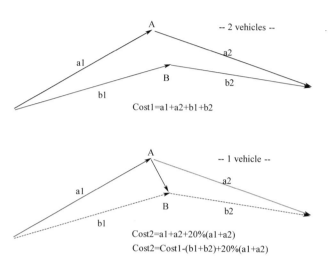

Fig. 1. Costs to collect two patients.

Considering now that we have one route that collects firstly patient A then performs a deviation to collect patient B and then goes to the destination. Once we use only one vehicle it is not necessary to pay b1 and b2, so these two terms are the brute saving. To make a deviation from A to B and then follow to destination, it is necessary to pay an extra fee. Nowadays in Portugal, the legislation establishes a percentage of the cost of this vehicle 20% (a1 + a2) per each additional patient collected. It is desirable to minimize the total cost of the transport service, mainly minimize the costs of individual transportation. In the same way, it is desirable to maximize the savings obtained with the shared transport of patients. This objective function follows the same principle of TOP objective function, that is maximize the total profits collected in the visited vertices in a graph. In this "real-world problem" is intended to maximize the total savings (compared with individual transportation) with a subset of patients that are selected to travel in the same vehicle.

2.1 Mathematical Model

Due to the importance of routing problems, the primary objective of a researcher is to obtain the optimal solution for a problem through linear integer programming models. In the case of the TOP, a set of problems is available in the literature, with mathematical formulation that is close to VRP formulation. Here we recall the formulation presented by Archetti et al. [5] for the TOP:

$$max \sum_{i \in V \setminus \{0\}} p_i \sum_{k=1}^{K} y_{ik} \tag{1}$$

$$\sum_{j \in V} x_{ijk} = y_{ik}, \forall i \in V, k = 1, \ldots, K \tag{2}$$

$$\sum_{j \in V} x_{ijk} = y_{ik}, \forall i \in V, k = 1, \ldots, K \tag{3}$$

$$\sum_{k=1}^{K} y_{0k} \leq K \tag{4}$$

$$\sum_{k=1}^{K} y_{ik} \leq 1, \forall i \in V \setminus \{0\} \tag{5}$$

$$\sum_{(i,j) \in \delta^+ (S)} x_{ijk} \geq y_{hk}, \ \forall S \subset V, \ 0 \in S, \ \forall h \in V(S), \ k = 1, \ldots, K \tag{6}$$

$$\sum_{(i,j) \in A} t_{ij} x_{ijk} \leq T_{max}, k = 1, \ldots, K \tag{7}$$

$$y_{ik} \in \{0, 1\}, \forall i \in V, k = 1, \ldots, K \tag{8}$$

$$x_{ijk} \in \{0, 1\}, \forall (i,j) \in A, k = 1, \ldots, K \tag{9}$$

In order to model the NEPT problem Oliveira et al. [6] established the following objective function:

$$max(\sum_{k=1}^{m} \sum_{j=2}^{n-1} \sum_{i=2}^{n-1} (c_{1j} + c_{jm}) x_{i,j}^k - \sum_{k=1}^{m} \sum_{j=2}^{n} \sum_{i=1}^{n-1} s(c_{1i} + c_{in}) TPS_{i,j}^k) \tag{10}$$

where the first parcel represents the saving costs when patient j is transported in a given route as a ride and the second parcel represents the additional payment to perform the deviation. The second parcel it is not linear and additional constraints were added to linearize the objective function. Also, the constraint (7) was replaced by three new constraints to model the route length established by the legislation.

The use of compact models of linear integer programming to solve the TOP is still limited to small instances of the problem in terms of the number of vertices considered. Recently, Oliveira et al. [6] reported that it is not possible to solve instances with more than one hundred patients with a fleet of ten ambulances with eight places each, using a compact model at the NEOS Server [7].

2.2   A Short Literature Review

The Team Orienteering Problem is NP-hard and for large instances it is not possible to obtain optimal solutions with a MILP model. The application of new methodologies, such as the branch and bound and price, offers the possibility to use the exact solution approach more efficiently. However, it is still a strategy not capable enough to deal with the majority of complex routing problems. Once the real problem can present an instance with several hundred patients and a fleet with dozens of vehicles it is necessary to provide a heuristic procedure to give solutions in a reasonable computational time.

The development of new heuristics to deal with complex optimization problems is huge. From Operational Research community and from Artificial Intelligence community recently arose powerful methods capable of dealing with large scale optimization. An alternative to the GA, we point out Particle Swarm Optimization (PSO) that gained interesting new contributions [8, 9].

A Gravitational Search Algorithm (GAS) is a new methodology based on concepts of gravitational law and mass interactions. GAS was proposed in 2009 by Rashedi et al. [10]. Since then an interesting number of works have been presented and we refer to Precup et al. [11] as an important contribution.

A very popular optimization method among research practitioners is the Evolutionary Algorithms/Genetic Algorithms (GAs) [12–15]. As an important feature in this methodology is signalled the fact that is a suitable tool to provide good quality solutions when the knowledge of the problem is somewhat limited.

3 The Methodology

As referred above in previous section TOP is a NP-hard problem and the well-known limitations of the exact procedures suggest us the use of approximate methodology. We opted to apply approximation methods, in particular GAs, to solve the presented problem. Our choice on using GAs was due to our experience acquired with these methods in previous works around the TOP.

As referred above in previous section TOP is a NP-hard problem and the well-known limitations of the exact procedures suggest to us the use of approximate methodology. We opted to apply approximation methods, in particular GAs, to solve the presented problem. Our choice of using GAs was due to our experience acquired with these methods in previous works around the TOP [13, 14], and also because it is easier to model the problem by adapting existing models we developed to this new variant. This "genetic algorithm's" characteristic justifies our choice to develop a heuristic procedure to deal with large instances.

We chose to develop a genetic algorithm using the JCell framework [8] that is available at the University of Malaga. To apply the framework to our problem we need to define the chromosome and the fitness function. Our choice was directed to a chromosome with an integer alphabet with a number of genes equal to the number of patients. Using the chromosome as a priority list, we establish an input to a constructive algorithm developed for the problem that returns valid solutions. We adopt the same strategy as the random key genetic algorithm [12] does to generate a solution from a chromosome. Each gene is related with one patient and gives us the priority of the patient to be included in the routes. The priority is defined by the allele value. Respecting the reordering of the alleles, we proceed with the construction of the routes, considering one patient at a time. With this type of constructive algorithm, we are able to obtain a valid solution for the problem, respecting all constraints. When a patient in its turn does not fit in any route, it will be abandoned, and we move on to the next patient. For those patients that can be transported, our algorithm finds the best position insertion among all available routes.

3.1   GAs on Routing Problems

Similar to other studies on routing problems, such as VRP, the TOP is an NP-hard problem. For many instances, it is not possible to achieve optimal solutions in a very short computing time. An alternative is the use of heuristic methods such as the genetic algorithms that can provide good solutions in a reasonable amount of time.

The GA is a search heuristic that imitates the natural process of evolution, as it is a process that is believed to occur in all species of living beings. This method uses nature-inspired techniques such as mutation, crossover, inheritance and selection, to generate solutions for optimization problems. The success of a GA depends on the type of problem to which the algorithm is applied in addition to its complexity.

In a GA, the chromosomes (or individuals) are represented as strings (vectors) that encode solutions for the problem, and later evolve toward better solutions by the means of an evolutionary process. This evolutionary process is responsible for executing crossovers and mutations within the population of chromosomes. The GA is able to control the evolutionary process and ensure the validation of chromosomes according to the limitations imposed by the problem (TOP). Designing a GA requires a genetic representation of the solution domain, as well as a fitness function to evaluate the solutions produced. The fitness function is able to determine the value of the objective function that is achieved by the routes encoded in a chromosome. In the GA, at each new generation (iteration), a renewed population of chromosomes is obtained, or in other words, a new set of encoded solutions is determined. Once decoded, each solution contains a set of routes, and each route includes a sequence of patients (collection points) to be visited under a given time limit. An example of a feasible solution is showed in Fig. 2. There are 13 collection points denoted as numbered vertices and

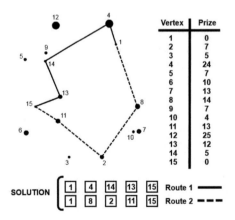

Fig. 2. Representation of a feasible solution for the TOP.

two vehicles are available (two routes). There is also a starting point (vertex 1) and an ending point (vertex 15). In this figure, the higher the prize of a vertex, the larger is the circle representing the vertex. In this case, not all vertices are included in the routes because otherwise it would make the solution infeasible by violating the time constraint. When applying our GA in the TOP case, given a graph with N vertices, a chromosome composed of $N - 2$ genes can be used to represent a solution for the optimization problem. Each gene is associated to a vertex in the graph. In order to always obtain feasible solutions, the allele of a gene contains a value that represents the priority of that vertex to be included in the routes (solution). Figure 3 contains a

GENE (vertex)	2	3	4	5	6	7	8	9	10	11	12	13	14
ALLELE	4	1	11	5	10	2	6	13	7	3	8	12	9

Fig. 3. Priority based chromosome.

representation of a chromosome in a priority based scheme for an instance with 15 vertices. The vertices 1 and 15 are respectively the starting and ending points, and therefore are omitted in the chromosome structure.

While addressing the TOP, the fitness function of the algorithm was set to correspond to the sum of all collected prizes in each visited vertex during the routes. Therefore, it calculates the total profit obtained in a certain solution (set of routes). Recalling the routing problem in the context of NEPT collection, the prize in each *patient* corresponds to the cost saved by avoiding individual transportation.

The constructive algorithm in the GA generates routes by interpreting chromosomes, and is able to handle all the constraints considered in the routing problem. For the TOP, the main constraint is on the route's duration. In order to assemble a route, customers (vertices) are added in from an availability list that is common to all routes. The list is obtained from a chromosome, by ordering the vertices following their priority levels. The attempt to add a customer to a route is only successful if that addition keeps that route feasible while not exceeding the given time limit. Once assigned to a route, a customer is then removed from the list and marked as checked; this way, the constraint stating that a vertex can only be visited once and by only one vehicle is always fulfilled. Each and every customer in the list is tested for an insertion into the routes of a solution. A chromosome must translate to as many routes as the number of vehicles available in a chosen TOP instance.

3.2   Three GAs Proposed for the TOP

In the first stage of this study, we developed three genetic algorithms (GAs) solely for the TOP, and each one embeds a constructive algorithm that differs from the others. We assessed their performance by executing computational tests on public benchmark instances and then we compared the results obtained with each GA and matched them against the results achieved by other state-of-the-art techniques presented in the literature. The three constructive algorithms for the different GAs are presented in the following subsections.

The GATOP1 Model. In this GA, the constructive algorithm is the less complex among the three developed. It processes a group of routes in a sequential way, one after the other. Following the list of available vertices, each vertex is checked for the possibility of being inserted in the route that is being processed at a given time. If a vertex is tested for an insertion and it is not added to the route, the next vertex on the list is the one that is then tested. As stated previously, this is a sequential process, where the route m + 1 only begins being assembled once route m is completed (no more insertions are possible).

The GATOP2 Model. In the GATOP2 model, following the availability list, each vertex is verified for the possibility of being assigned to the set of routes being assembled at a given time. This way, each vertex is tested for an insertion in the route m + 1 if it was not possible to assign it to route m. So, the construction of routes is done in a mixed procedure, which is sequential, yet it also includes parallel processing.

The GATOP3 Model. The model named GATOP3 incorporates in its GA a constructive algorithm that performs a parallel processing while assembling the routes. In this case, a vertex is tested in all routes in the hope of finding the best possible insertion among the m routes being assembled at a given time. That happens when the vertex is in the route, and positioned within that route, where it adds less costs (travel time spent) to that route when compared to the others. Once the best fit is found, the vertex is assigned and the next one on the list is tested for possible insertions.

4 Computational Experiments and Test Results

Several experiments were conducted in order to evaluate the performance of our GAs. First, we chose 24 TOP instances from the sets published by Chao [16] in order to assess which genetic algorithm performed best. The first phase of tests focused on running the three GAs on the set of 24 TOP instances. We chose them by taking into consideration their complexity, and thus we have selected both, a few easy instances and instances of a medium-to-hard level of difficulty to solve. We tuned each GA to meet their best-known performance configuration. A summary of the results obtained during these tests is given in Table 1, where the average values of 10 runs per instance are showed for each GA (models GATOP1, GATOP2 and GATOP3).

Table 1. Results obtained during the first phase of tests.

Instance	Fitness level (Score)									Best score
	GATOP-1			GATOP-2			GATOP-3			
	Avg	fmin	fmax	Avg	fmin	fmax	Avg	fmin	fmax	
p4.2f	609,6	565	643	662,0	648	684	685,2	681	**687**	687
p4.2.o	1004,0	923	1081	1116,7	1048	1178	1169,8	1027	1208	**1218**
p4.3.j	764,8	722	818	826,9	790	854	855,2	851	**861**	861
p4.3.p	985,7	935	1062	1119,9	1036	1174	1194	1174	1205	**1222**
p4.4.k	736,7	697	775	794,5	733	821	820,6	820	**821**	821
p4.4.r	992,6	922	1054	1139,0	1110	1171	1182,6	1174	1191	**1211**
p5.2.p	1066,5	1045	1090	1090,0	1055	**1150**	1150	1150	**1150**	1150
p5.2.z	1627,0	1515	1675	1633,0	1595	1660	1680	1680	**1680**	1680
p5.3.t	1225,0	1205	1245	1242,5	1220	**1260**	1259	1255	**1260**	1260
p5.3.y	1500,5	1455	1555	1561,0	1520	1590	1590	1590	1590	**1595**
p5.4.w	1303,5	1250	1355	1357,0	1310	1380	1380	1380	1380	**1390**
p5.4.z	1501,0	1460	1535	1525,5	1455	1575	1291,6	158	1585	**1620**
p6.2.l	1093,2	1050	1110	1090,2	1068	**1116**	1116	1116	**1116**	1116
p6.2.m	1161,6	1146	**1188**	1157,4	1134	1176	1188	1188	**1188**	1188
p6.3.m	1033,8	996	**1080**	1069,8	1056	**1080**	1680	1680	**1680**	1188
p6.3.n	1124,4	1104	1158	1158,6	1146	**1170**	1170	1170	**1170**	1170
p6.4.m	885,0	852	906	902,4	888	**912**	912	912	**912**	912
p6.4.n	1051,8	1032	**1068**	1068,0	10668	**1068**	1068	1068	**1068**	1068
p7.2.l	711,3	692	722	740,8	692	763	759,8	751	**767**	767
p7.2.s	967,1	910	1030	1071,3	991	1110	1122,8	1114	1135	**1136**
p7.3.o	766,6	729	808	845,9	812	865	866,8	858	**874**	874
p7.3.t	947,8	891	1017	1072,5	1037	1112	1115	1111	**1117**	1116
p7.4.p	776,7	755	814	820,8	795	844	830,8	814	842	**846**
p7.4.s	896,6	855	958	991,6	929	1020	1017,2	1006	**1022**	1022

Algorithm 1 • Proposed Constructive Algorithm GATOP3

```
input: TOP data, chromosome,s,e \\s is start-path, e is
end-path
P₀ \\P is the set of points to collect
A₀ ≡ P₀ \\A is the set of points available to collect
V₀ = Ø \\V is the set of visited points
begin
  for k ← 1 to numRoutes do
  \\Initialize Route
  Rᵏ : s → e
end for
for j ← 1 to |A| do
  i* ← argmin{chromosomeᵢ |Aⱼ }
  Aⱼ ← Aⱼ₋₁\{i*}
  insertCost* = ∞; k* = NULL; j* = NULL;
  for k ← 1 to numRoutes do
    for j ← 1 to |Rᵏ| − 1 do
      if i* fits in Rᵏ in jᵗʰ position then
        if insertCost < insertCost* then
          insertCost* = insertCost; k* = k; j* = j;
        end if
      end if
    end for
  end for
  if k* ≠ NULL then
    Rᵏ*← i*\\i* is placed in j*ᵗʰ position
    Vⱼ ← Vⱼ₋₁ U {i*}
  end if
end for
TotalProfit = Σᵢₑᵥ pᵢ pᵢ
end
output: a valid solution
```

In terms of parameter configuration for the GA, the following values were adopted:

- Population size: $2 \times N$, with N being the number of vertices in a graph.
- Iterations: 1000
- Crossover Rate: $N \div 2$
- Mutation Rate (per gene): $1 \div N$

We test this JCell based genetic algorithm with large size instances that represent the real problem. It was not possible to obtain optimal value for these instances. Nevertheless, NEOS Server can solve the large instances using only one vehicle

(Orienteering Problem - OP). Considering this situation, we developed a heuristic procedure to solve the TOP that is based in successive OP solutions. Iteratively, to the remaining unvisited vertices, we solve the problem using the compact formulation with one vehicle. Table 2 compares the results obtained with NEOS Server ("NeosS" line) and GA heuristic ("heur" line) in a set of 30 instances, considering real data, with vehicles with capacity of 4 (left side), and with vehicles with capacity of 8 (right side). In general, the GATOP3 produce good values, and we point out that GATOP3 outperforms NEOS Server heuristic, because GATOP3 has a better value in 22 of 30 instances, has equal value in 4 instances, and worst value in 4 instances.

Table 2. Results obtained with real data

nodes		vehicles x capacity			nodes		vehicles x capacity		
		3x4	7x4	10x4			3x8	7x8	10x8
20	NeosS	247	208	208	20	NeosS	226	210	210
	heur	**247**	**196**	225		heur	**226**	**196**	225
40	NeosS	582	421	335	40	NeosS	490	355	355
	heur	**579**	**420**	**348**		heur	**476**	**334**	**321**
60	NeosS	991	790	679	60	NeosS	860	619	535
	heur	**976**	**773**	**661**		heur	**811**	**552**	**466**
80	NeosS	1365	1157	1017	80	NeosS	1205	882	729
	heur	**1359**	**1146**	**1017**		heur	**1152**	**832**	**661**
100	NeosS	1737	1518	1368	100	NeosS	1546	1164	944
	heur	**1137**	**1055**	1373		heur	**1500**	1185	**928**

5   Conclusions

This paper presents a variant of Team Orienteering Problem that models the non-emergency patient transport service. This service is provided by the National Health Care and the main objective of this study is to present a methodology to reduce the service costs. In order to produce solutions to a real-world largescale problem a genetic algorithm has been developed using the JCell framework. The computational results obtained guarantee the good quality of the solutions, and the cost savings are significant. We presented three genetic algorithms (GAs) that were developed to solve the TOP, each one with a different constructive procedure. Some computational experiments are presented and the results achieved have shown the performance of the techniques adopted. The GA will be integrated in a Decision Support System that could be used to aid in the generation of routes in a real life process of patients transport service. From our experiments with the GATOP3 we concluded that it is more profitable and outperforms GATOP1 and GATOP2.

Acknowledgments. This work was funded by the "Programa Operacional Fatores de Competitividade – COMPETE" and by the FCT - Fundação para a Ciência e Tecnologia in the scope of the project: FCOMP-01-0124-FEDER-022674.

The authors would like to thank the anonymous reviewers for their valuable comments and suggestions to improve the paper. The authors would like to thank Dr. Elaine DeBock for reviewing the manuscript.

References

1. DRE: Portaria n.º 142-A/2012. Diário da República, 1.ª série — N.º 94 — 15 de maio de 2012, 2532-(2)-2532-(3) (2012)
2. DRE: Portaria n.º 142-B/2012. Diário da República, 1.ª série — N.º 94 — 15 de maio de 2012, 2532-(3)-2532-(6) (2012)
3. Vaisblat, A., Albert, D.: Medical non-emergency patient centered scheduling solution. New Magenta Papers, 40 (2013)
4. Vansteenwegen, P., Souffriau, W., Oudheusden, D.V.: The orienteering problem: a survey. Eur. J. Oper. Res. **209**(1), 1–10 (2011)
5. Archetti, C., Speranza, M.G., Vigo, D.: Vehicle routing problems with profits. Working paper (2012)
6. Oliveira, J.A., Ferreira, J., Dias, L., Figueiredo, M., Pereira, G.: Non emergency patients transport - a mixed integer linear programming. In: 4th International Conference on Operations Research and Enterprise Systems (ICORES 2015), Lisboa, Portugal, pp. 262–269 (2015). ISBN 978-989-758-075-8
7. Czyzyk, J., Mesnier, M.P., Moré, J.J.: The NEOS server. IEEE J. Comput. Sci. Eng. **5**, 68–75 (1998)
8. Valdez, F., Melin, P., Castillo, O.: An improved evolutionary method with fuzzy logic for combining particle swarm optimization and genetic algorithms. Appl. Soft Comput. **11**(2), 2625–2632 (2011)
9. El-Hefnawy, N.A.: Solving bi-level problems using modified particle swarm optimization algorithm. Int. J. Artif. Intell. **12**(2), 88101 (2014)
10. Rashedi, E., Nezamabadi-Pour, H., Saryazdi, S.: GSA: a gravitational search algorithm. Inf. Sci. **179**(13), 2232–2248 (2009)
11. Precup, R.E., David, R.C., Petriu, E.M., Preitl, S., Paul, A.S.: Gravitational search algorithm-based tuning of fuzzy control systems with a reduced parametric sensitivity. In: Soft Computing in Industrial Applications, pp. 141–150. Springer, Heidelberg (2011)
12. Bean, J.C.: Genetics and random keys for sequencing and optimization. ORSA J. Comput. **6**, 154–160 (1994)
13. Mota, G., Abreu, M., Quintas, A., Ferreira, J., Dias, L.S., Pereira, G.A., Oliveira, J.A.: A genetic algorithm for the TOPdTW at operating rooms. In: Lecture Notes in Computer Science, vol. 7971, pp. 304–317 (2013)
14. Ferreira, J., Quintas, A., Oliveira, J.A., Pereira, G., Dias, L.: Solving the team orienteering problem: Developing a solution tool using a genetic algorithm approach. In: Soft Computing in Industrial Applications: Advances in Intelligent Systems and Computing, vol. 223, pp 365–375 (2014)
15. Alba, E., Dorronsoro, B.: Software for cGAs: the JCell framework. In: Cellular Genetic Algorithms, pp. 153–163. Springer, US (2008)
16. Chao, I., Golden, B.L., Wasil, E.A.: The team orienteering problem. Eur. J. Oper. Res. **88**, 464–474 (1996)

Modeling of Flow Shop Scheduling with Effective Training Algorithms-Based Neural Networks

Dipak Laha[✉] and Arindam Majumder

Department of Mechanical Engineering, Jadavpur University, Kolkata, India
dipaklaha_jume@yahoo.com, arindam2012@gmail.com

Abstract. This paper deals with the performance comparison of three most effective neural network backpropagation training algorithms such as gradient descent, Boyden, Fletcher, Goldfarb and Shanno (BFGS) based Quasi-Newton (Q-N) and Levenberg-Marquardt (L-M) algorithms. The training of the neural network is carried out based on random datasets considering optimal job sequences of the permutation flow shop problems. In the present investigation, a goal of 0.001 of MSE or 3000 of epochs is set as a goal of learning. The overfitting and overtraining are not allowed during model building to avoid poor generalization ability. The performance of different learning techniques is reported in terms of both solution quality and computational times. The computational results demonstrate that the L-M performs best among the three algorithms with respect to both MSE and R^2. However, the gradient descent algorithm is the fastest among them.

Keywords: Scheduling · Flow shops · Makespan · ANN
Gradient descent algorithm · Q-N algorithm · L–M algorithm

1 Introduction

The flow shop scheduling, one of the earliest scheduling system encountered, is one of the most versatile and widely used systems. Since a larger percentage of engineering products are processed in flow shop environment, this scheduling has led to the development of some specialized scheduling systems like no-wait flow shop, hybrid or flexible flow shop or multi-stage scheduling, blocking flow shop, and cellular manufacturing system.

In flow shop scheduling, each of n jobs from a set $N = \{1, 2 \ldots, n\}$ is processed on a set $M = \{1, 2, \ldots, m\}$ of m machines in the same technological order as given by the indexing of machines with the objective of optimizing some criterion. Setup times are independent of the processing times and are included in the operation times. Each machine can perform one job at a time and preemption is not allowed.

Since flow shop scheduling problems belong to the class of NP-hard, approximation optimization methods such as heuristics and metaheuristics are mostly preferred to solve these problems, especially large problem sizes. Noteworthy heuristics in flow shop scheduling on makespan criterion have been developed by Johnson [1], Palmer

© Springer Nature Switzerland AG 2019
B. K. Ane et al. (Eds.): WSC 2014, AISC 864, pp. 107–117, 2019.
https://doi.org/10.1007/978-3-030-00612-9_10

[2], Campbell, Dudek and Smith [3], Gupta [4], Dannenbring [5], Nawaz, Enscore and Ham [6], Rajendran [7], Koulamas [8] and Suliman [9].

Recently, for modelling NP-hard flow shop scheduling problems, artificial neural networks (ANN) have been successfully used due to its good learning ability. Haq, Ramanan, Shashikant and Sridharan [10] proposed a hybrid approach by combining ANN with GA to find the job sequence for permutation flow shop scheduling to minimize the makespan. El-bouri, Subramaniam and Popplewell [11] presented an approach to enhance local search in permutation flow shop by constructing initial sequences by training ANN. Ramanan, Sridharan, Shashikant and Haq [12] have made an attempt to find a job sequence in the parallel flow shop scheduling with minimized makespan using ANN to obtain initial sequences for heuristic proposed by Suliman [9] and GA. Akyo [13] employed ANN to model six different heuristic algorithms applied to the n-job, m machine flow shop scheduling problem with minimization of makespan. Lee and Shaw [14] developed a twolevel neural network that incrementally learns sequencing knowledge based on the knowledge gained from a set of existing training exemplars and can sequence a set of jobs on a real-time basis based on ANN to develop hybrid GA and the results showed an improvement in solution quality and computational time as compare to the conventional GA.

In this study, an attempt has been made to investigate a comparative study on the effectiveness of three popular learning algorithms such as gradient descent with adaptive learning, BFGS updated Q-N and L-M algorithms on the performance of backpropagation neural networks by modelling a permutation flow shop scheduling considering the objective of minimizing makespan.

2   Artificial Neural Network

The ANN is a computational model inspired by the biological nerve system and it has a very good learning ability between input and output patterns of a complex system. The feed-forward backpropagation neural network (BPNN) is a very popular model.

There are some noteworthy research contribution in the development of heuristic rules in order to improve the performance of the BPNN. Jacobs [15] proposed three heuristics such as momentum procedure, delta-bar-delta procedure and hybrid of momentum and delta-bar-delta rules and compared with the steepest decent method. Later on, the algorithm of Jacobs [15] was extended by Minai and Williams [16] and its superior performance was shown over the conventional BPNN [17]. Liang, Liu, Li and Yuan [18] introduced a new optimal step-size algorithm in BPNN for accelerating convergence speed of the network. Tollenaere [19] presented a new adaptive acceleration strategy called super self-adaptive backpropagation to make the ANN training faster. During this study the proposed approach was compared with the existing approach and it was found that it can converge faster than the original back propagation algorithm with slight instability. Sarkar [20] investigated several existing modifications on learning rate rule to speed up the backpropagation learning algorithm. During his study, some of these modifications were combined together to get faster convergence rate. The conclusion of this study revealed that every modification, except self-determination of adaptive learning rate, requires a good learning rate coefficient to be supplied by the user.

The most popular learning algorithm for the training of BPNN is gradient descent algorithm. Due to slow convergence to a local minimum, some improved learning algorithms have been found effective for training the BPNN are available in the literature [21–24]. Gradient descent algorithm has been improved by introducing adaptive learning rate. Similarly, Q-N algorithm has been updated by Boyden, Fletcher, Goldfarb and Shanno [27] which is known as BFGS updated Q-N algorithm. The L-M algorithm [23] has shown superior performance, convergence and predicting values in many experimental studies. A brief contribution of these algorithms are described below.

2.1 Gradient Descent Algorithm

The major disadvantage of the gradient descent algorithm is its constant learning rate throughout training. It is due to the fact that the performance of the algorithm is very sensitive to the proper setting of the learning rate and in practical it is very difficult to determine the optimal setting for the learning rate before training. If the learning rate is high, the algorithm can oscillate and become unstable. On the other hand, if the learning rate is small, it takes too long time for convergence. However, this can be improved, if the learning rate is allowed to change with iteration.

The modified version of this is known as gradient decent algorithm with adaptive learning rate.

Adaptive learning rate rule [25] is a new efficient learning scheme, which performs a direct adaptation of the weight step based on local gradient information. In this approach, for each weight the individual update-value δij is introduced, which exclusively determines the size of the weight-update. The individual update-value (δij) is developed during the learning process and it is based on its local sight on the error function E and is defined as:

$$\delta_{ij}^{(t)} = \begin{cases} \eta^+ \times \delta_{ij}^{(t-1)}, \text{if } \frac{\partial E^{(t-1)}}{\partial w_{ij}} \times \frac{\partial E^{(t)}}{\partial E^{(t-1)}} > 0 \\ \eta^- \times \delta_{ij}^{(t-1)}, \text{if } \frac{\partial E^{(t-1)}}{\partial w_{ij}} \times \frac{\partial E^{(t)}}{\partial w_{ij}} < 0 \\ \delta_{ij}^{(t-1)}, \text{else} \end{cases} \quad (1)$$

After obtaining the update-value of each weight, the weight updates itself following a simple rule as shown below:

$$\delta w_{ij}^{(t)} = \begin{cases} -\delta_{ij}^{(t)}, \text{if } \frac{\partial E^{(t)}}{\partial w_{ij}} > 0 \\ +\delta_{ij}^{(t)}, \text{if } \frac{\partial E^{(t)}}{\partial w_{ij}} < 0 \\ 0, \text{else} \end{cases} \quad (2)$$

$$w_{ij}^{(t+1)} = w_{ij}^{(t)} + \delta w_{ij}^{(t)} \tag{3}$$

The parameters of the gradient descent algorithm are considered as: initial learning rate = 0.01; η = 1.2 and η = 0.5.

2.2 Quasi-newton Algorithm

Q-N learning algorithm is an alternative to conjugated gradient descent algorithm for fast convergence in training neural network. It is somewhat similar to the Newton's method with line search. However, the basic difference lies on the iterative approximation and updation of inverse Hessian. In Newton's method the inverse of Hessian matrix is generated by computing second order partial derivatives of error function. However, computation of the inverse of Hessian matrix for feed-forward neural networks is very difficult and time consuming. The Q-N method avoids such computing by considering an approximation to the Hessian which updates iteratively by a low rank updation. This makes Q-N learning algorithm easy for computation.

The Q-N method is first developed by Davidson, Fletcher and Powell (known as DFP method) [26]. But it was soon replaced by Broyden, Fletcher, Goldfarb and Shanno (BFGS) [27–30], which is considered as the most efficient Q-N method. The chain rule used to update weight iteratively during this method is as follows:

- compute quasi-Newton direction $\partial w = H_k \partial_{gk}$
- determine step size t (e.g., by backtracking line search)
- compute $w_{k+1} = w_k + \alpha \partial w$
- compute H_{k+1}.

The initial parameters of BFGS updated Q-N learning algorithm are taken as: maximum step size = 26; minimum step length = 10^{-6}; maximum step length = 100; upper limit on change in step size = 0.5; lower limit on change in step size = 0.1; initial step size in interval location step = 0.01; scale factor that determines sufficiently large step size = 0.1 and scale factor that determines sufficient reduction in performance = 0.001.

2.3 Levenberg-Marquardt Algorithm

L-M algorithm is a technique to locate the minimal of a multivariate function which is articulated as the sum of squares of non-linear real-valued functions. This method operates with a combinational advantage of gradient descent and the Gauss-Newton method. If the current solution is far away from the required one then the algorithm behaves like a gradient descent method with slow, but guaranteed convergence. On the other hand, when the current solution is near to the correct solution, it becomes a

Gauss-Newton method. The mathematical representation used to upgrade layer weight in L-M algorithm [31] is as follows:

$$w_{k+1} = w_k - (J_k^T J_k + \lambda I)^{-1} J_k^T E \qquad (4)$$

Where, I is the Identical matrix and J is the Jacobian of n output errors with respect to m weight of the neural network. The learning parameter (λ) adjusts automatically at each iteration in order to make convergence consistent.

If $\lambda = 0$ the algorithm becomes Gauss-Newton method and for large value of λ, it becomes the gradient descent algorithm. The parameters of the L-M algorithm are selected as: minimum performance gradient $= 10^{-6}$; initial $\lambda = 0.001$; λ decrease factor $= 0.1$; λ increase factor $= 10$ and maximum $\lambda = 10^{10}$.

3 Modeling Flow Shop Scheduling

In the present study, the proposed neural networks for n-job, m-machine flow shop scheduling are trained using the existing dataset. The different steps used to train these neural networks are as follows:

Step 1: It deals with determining neural network architecture. The neural network considered consists of three layers such as one input layer, one hidden layer and one output layer. The number of nodes introduced in the input layer of these proposed neural networks is three times the number of the machines and are allocated as follows:

- The first m nodes contain job's processing times on each of the m machines
- The middle m nodes contain the average processing time requirements on each of the m machines
- The last m nodes contain the standard deviation of the processing times on each of the m machines.

On the other hand, the hidden layer contains h nodes and are determined by using parametric study process. While only one node is considered in the output layer. The output node assumes value between 0.1 and 0.9 to accommodate the sigmoidal function which is used in these training algorithms.

Step 2: A sufficiently representative set of data is prepared to train the proposed neural network. The values for the input-output patterns must be between zero and one. Hence, the input data must be normalized. The minimum and maximum processing times in the flow shops considered in this study are 1 and 100 time units. The equation used to compute the input nodal value for i-th job is presented as follows:

$$node(q) = \overline{p} \begin{cases} \frac{Pi,q}{100} \\ \frac{(q-m)}{100} \\ \sqrt{\frac{X(q-2m)^{-np^2}(q-2m)}{(n-1)\times 10^4}} \end{cases} \qquad (5)$$

$$\text{Where, } \overline{p(k)} = \frac{1}{n}\sum\nolimits_{i=1}^{n} P_i, k \text{ and } X_{(k)} = \sum\nolimits_{i=1}^{n} p^2 i, k \tag{6}$$

The target output (O[i]) for the i-th job in the optimal sequence is determined by [10, 11]:

$$O_{[i]} = 0.1 + \frac{08([i] - 1)}{n - 1} \tag{7}$$

Step 3: The training patterns are then used to train the neural networks given in Step 1. Then these trained neural networks are used to solve new flow shop scheduling problem.

4 Computational Results

In this section, we investigate a comparative study on the effectiveness of three popular algorithms such as gradient descent with adaptive learning, BFGS updated Q-N and L-M algorithms on the performance of backpropagation neural networks by modeling a permutation flow shop scheduling considering the objective of minimizing makespan. In this paper, above algorithms were coded in MATLAB 2009a and executed on a PC with Intel i52450 M CPU with 4 GB RAM running at 2.50 GHZ.

Following the similar experimental framework given by Haq et al. [10], the training dataset for each neural network was constructed by generating randomly problems of 400, 500 and 715 considering the problem size with number of jobs (n) = 5, 6, 7 and number of machines (m) = 5 respectively. Each of these problems is solved by complete enumeration to obtain optimal sequence of jobs which is used to set the target output. As a result, a total of 10005 input-output pairs called exemplars are generated as training dataset. In order to evaluate the predictive quality of these neural networks, three performances indices, popular in the neural network literature, are used in the present experimentation: mean square error (MSE), R^2 coefficient and CPU time/epoch.

The stopping criteria set for these three networks are given in Table 1. Also, the appropriate number of nodes in the hidden layer obtained by parametric study process is shown in Figs. 1, 2 and 3. It is seen from the figures that the optimum number of hidden neurons for each of the gradient descent with adaptive learning rate, BFGS updated QN and L–M algorithms based propagation neural networks are 250, 150 and 300 respectively. Each of these three training algorithms was run five times based on the respective stopping criterion as given in Table 1. Figure 4 shows the convergence characteristics of these three neural networks for a particular instance.

Table 1. Stopping criteria for the learning techniques

Criterion	GD with adaptive learning rate	BFGS based Q-N	L–M
MSE	0.001	0.001	0.001
Epoch	3000	1000	500
Max validation failure	5	5	5

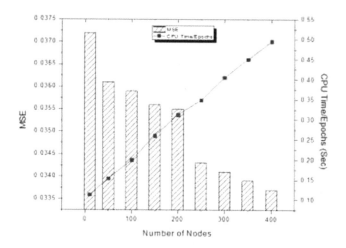

Fig. 1. Comparison of training time (in CPU time/epochs) and MSE for various number of hidden nodes of GD with adaptive learning rate backpropagation NN

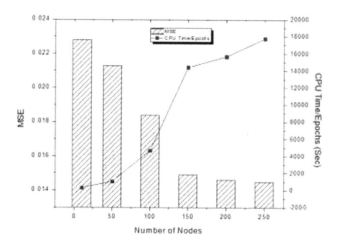

Fig. 2. Comparison of training time (in CPU time/epochs) and MSE for various number of hidden nodes of BFGS Q-N backpropagation NN

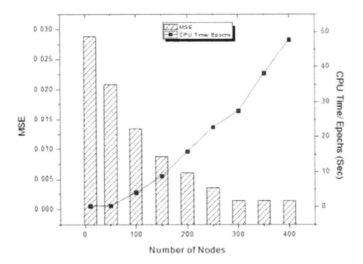

Fig. 3. Comparison of training time (in CPU time/epoch) and MSE for various number of hidden nodes of L–M algorithm

Table 2 shows the comparative performance measures of these three neural networks. From the results of the table, it is found that the L–M based neural network in terms of MSE and R^2 performs best among all the algorithms. It is also observed that the number of epochs taken by L–M algorithm (500 epochs) for convergence is less than the number of iteration taken by gradient descent with adaptive learning rate (3000 epochs) and BFGS updated Q–N algorithms (1000 epochs). These tendencies are expected due to the fact that at higher error value LevenbergMarquardt algorithm becomes gradient descent method with small step size and it turns into Gauss-Newton method near minimum error. In this way, the convergence is always faster and never slowed down even at higher or lower error value. However, in gradient descent with adaptive learning rate, there is no need to calculate Jacobian or approximate Hessian matrix, it requires less computational time (CPU time) to complete one epoch than the other two learning techniques.

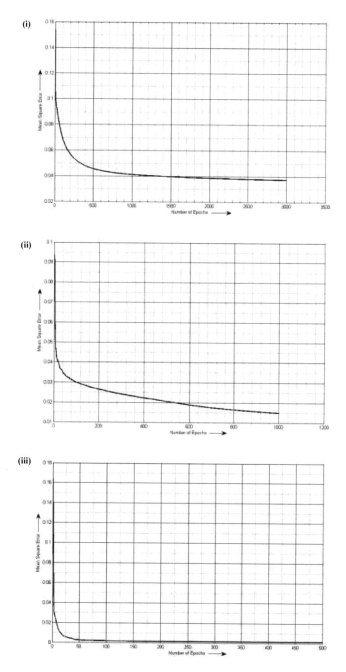

Fig. 4. Convergence characteristics of (i) GD with Adaptive learning rate backpropagation NN, (ii) BFGS Q-N backpropagation NN and (iii) L–M backpropagation NN

Table 2. Performance Comparison between GD with adaptive learning rate backpropagation NN, BFGS Q-N backpropagation NN and L–M backpropagation NN

Performance measure		Gradient descent	BFGS Q-N	L–M
MSE	Min	0.0381	0.0152	0.0023
	Max	0.0334	0.0149	0.0010
	Mean	0.0356	0.01505	0.00147
R^2	Min	70.52	82.07	99.17
	Max	73.87	80.71	98.89
	Mean	74.8	81.39	99.04
CPU time (s)	Min	0.351	14.382	28.03
	Max	0.346	14.355	27.37
	Mean	0.348	14.3685	27.56

5 Conclusions

In this paper, three effective backpropagation training algorithms, namely, gradient descent with adaptive learning rate, BFGS updated Quasi-Newton and Levenberg-Marquardt were presented to train the neural network for modelling permutation flowshop scheduling. The computational results show that the Levenberg-Marquardt performs best among the three algorithms with respect to both MSE and R^2. Levenberg-Marquardt algorithm takes less iteration for training of neural network as compare to the other algorithms while the adaptive learning rate based gradient descent algorithm requires less computational time than the other two learning techniques.

References

1. Johnson, S.: Optimal two- and three-stage production schedules with setup times included. Naval Res. Logistics Q. **1**(1), 61–68 (1954)
2. Palmer, D.: Sequencing jobs through a multi-stage process in the minimum total time – a quick method of obtaining a near optimum. Oper. Res. Q. **16**(1), 45–61 (1965)
3. Campbell, H.G., Dudek, R.A., Smith, M.L.: A heuristics Algorithm for the n job m machine sequencing problem. Manage. Sci. **16**(10), 630–637 (1970)
4. Gupta, J.N.D.: A functional heuristic algorithm for the flow shop scheduling problem. Oper. Res. Q. **22**(1), 39–47 (1971)
5. Dannenbring, D.: An evolution of flowshop scheduling heuristics. Manage. Sci. **23**(11), 1174–1182 (1977)
6. Nawaz, M., Enscore, J., Ham, I.: A Heuristic algorithm for the m-machine, n-job flowshop sequencing problem. Omega **11**(1), 91–95 (1983)
7. Rajendran, C.: Theory and methodology heuristics for scheduling in flow shop with multiple objectives. Eur. J. Oper. Res. **82**(3), 540–555 (1995)
8. Koulamas, C.: A new constructive heuristic for the flowshop scheduling problem. Eur. J. Oper. Res. Soc. **105**(1), 66–71 (1998)
9. Suliman, S.: A two-phase heuristic approach to the permutation flow-shop scheduling problem. Int. J. Prod. Econ. **64**(1–3), 143–152 (2000)

10. Haq, N.A., Ramanan, T.R., Shashikant, K.S., Sridharan, R.: A hybrid neural network–genetic algorithm approach for permutation flow shop scheduling. Int. J. Prod. Res. **48**(14), 4217–4231 (2010)

11. El-bouri, A., Subramaniam, B., Popplewell, N.: A neural network to enhance local search in the permutation flowshop. Comput. Ind. Eng. **49**(1), 182–196 (2005)

12. Ramanan, T.R., Sridharan, R., Shashikant, K.S., Haq, A.N.: An artificial neural network based heuristic for flow shop scheduling problems. J. Intell. Manuf. **22**(2), 279–288 (2011)

13. Akyol, D.: Application of neural networks to heuristic scheduling algorithms. Comput. Ind. Eng. **46**(4), 679–696 (2004)

14. Lee, I., Shaw, M.J.: A neural-net approach to real time flow shop sequencing. Comput. Ind. Eng. **38**(1), 125–147 (2000)

15. Jacobs, R.A.: Increased rates of convergence through learning rate adaptation. Neural Netw. **1**, 295307 (1998)

16. Minai, A.A., Williams, R.D.: Acceleration of back-propagation through learning rate and momentum adaptation. In: Proceedings of the International Joint Conference on Neural Networks, Washington, DC (1990a)

17. Minai, A.A., Williams, R.D.: Backpropagation heuristics: a study of the extended delta-bar-delta algorithm. In Proceedings of the International Joint Conference on Neural Networks, San Diego, CA (1990b)

18. Liang, G., Liu, C., Li, Y., Yuan, F.: Training feed-forward neural networks using the gradient descent method with the optimal stepsize. J. Comput. Inf. Syst. **8**, 1359–1371 (2012)

19. Tollenaere, T.: SuperSAB: fast adaptive back propagation with good scaling properties. Neural Netw. **3**(5), 561–573 (1990)

20. Sarkar, D.: Methods to speed up error back-propagation learning algorithm. ACM Comput. Surv. (CSUR) **27**, 519–544 (1995)

21. Andersen, T.J., Wilamowski, B.M.A.: Modified regression algorithm for fast one layer neural network training. World Congr. Neural Netw. **1**, 687–690 (1995)

22. Battiti, R.: First-and second-order methods for learning: between steepest descent and Newton's method. Neural Comput. **4**(2), 141–166 (1995)

23. Hagan, M.T., Menhaj, M.B.: Training feedforward networks with the Marquardt algorithm. IEEE Trans. Neural Netw. **5**(6), 989–993 (1994)

24. Shah, S., Palmieri, F.: MEKA-A fast, local algorithm for training feedforward neural networks. Neural Netw., 41–46 (1990)

25. Gavin, H.: The Levenberg-Marquardt method for nonlinear least squares curve-fitting problems. Department of Civil and Environmental Engineering, Duke University (2011)

26. Fletcher, R.: Practical Methods of Optimization. Wiley (2013)

27. Broyden, C.G.: The convergence of a class of double-rank minimization algorithms general considerations. IMA J. Appl. Math. **6**(1), 76–90 (1970)

28. Goldfarb, D.: A family of variable-metric methods derived by variational means. Math. Comput. **24**(109), 23–26 (1970)

29. Shanno, D.F.: Conditioning of quasi-Newton methods for function minimization. Math. Comput. **24**(111), 647–656 (1970)

30. Riedmiller, M., Braun, H.: A direct adaptive method for faster backpropagation learning: The RPROP algorithm. Neural Netw., 586–591 (1993)

Intelligent Land Cover Detection in Multi-sensor Satellite Images

R. Jenice Aroma$^{(\boxtimes)}$ and Kumudha Raimond

Department of Computer Science and Engineering, Karunya University,
Chennai, India
jenicearoma@karunya.edu.in, kraimond@karunya.edu

Abstract. Nowadays, the availability of satellite images is higher due to the launch of more number of earth observation and inter-planetary mission satellites. The satellite images are widely used in different early warning, risk assessment and disaster models. Hence, the need for efficient interpretation of these resources is also high. Generally, it is a primary issue in satellite image analysis to detect the specific region or individual objects in multiple sensor satellite images of varied spatial resolution. If traditional classifiers are used for such object detection, it needs much training time and tedious ground truth labeling. Hence, the proposed model is focused on reducing the above said complexities and offer efficient detection of the chosen land cover region using Speeded Up Robust Features (SURF), a widely used robust local feature descriptor. The performance of SURF has been evaluated using three different sensor images of moderate resolution for a common study region.

Keywords: Object detection · Satellite image · Multiple sensor
SURF

1 Introduction

The adverse climate changes all over the world leads to high level of environmental degradation and it brings the need for effective monitoring of the natural ecosystems like Terrains, coastal areas, crop lands, forests and oceans. To build an efficient hazard mapping models, the field measurements and other related ground sensor measurements are insufficient. It leads to the wide usage of satellite images in hazard prediction. In general, the satellite images are characterized based on their spatial, spectral and temporal resolutions [1]. The surveillance and urban development models require high spatial resolution for locating individual objects [2]. Similarly, high temporal resolution images are used for extraction of climate parameters to monitor environmental changes and to analyze the historical change trend [3–5]. The satellite image undergoes tedious preprocessing step to rectify the atmospheric and geometric corrections which may consume much time but it would be favorable for efficient image interpretation. Thus need for advanced imaging and interpretation techniques with reduced time and increased accuracy is a major need in satellite image analysis [6, 7].

The satellite image based applications are commonly applied for spatial monitoring and land cover change mapping which needs an advanced level of object detection

© Springer Nature Switzerland AG 2019
B. K. Ane et al. (Eds.): WSC 2014, AISC 864, pp. 118–128, 2019.
https://doi.org/10.1007/978-3-030-00612-9_11

from satellite images. The object detection algorithms in general, can be categorized into two different groups as global features and local features based algorithms. The global features are extracted from the homogenous regions of the segmented objects of an image but the local features are highly robust features which could overcome the variations on illumination and different viewpoints [8]. In local features, the salient feature points are initially detected and a region around those points are constructed with respect to geometrical and illumination variations to extract these features like Scale Invariant Feature Transform (SIFT), Harris features and SURF etc.

In case of satellite image based disaster support models, the object detection must be done at a reduced time whereas the tedious training and labeling may delay the timely detection. It needs automated image analysis based on robust local features for increased accuracy and faster computation [9]. In traditional image classification approaches for spatial analysis the color, texture and geometrical features are extracted and a trained classifier is then used for accurate recognition of individual objects from spatial data. It has been inferred from the literature that the integration of multiple features extracted using machine learning methods could achieve better accuracy [10]. But these learning frameworks are still not much efficient in large scale data analysis due to high time consumption for training and insufficient ground truth labeled data. To resolve these limits SURF based object detector is introduced in 2013 to improve training efficiency [11]. Also, the entire spatial data of a specified region cannot be covered by a single sensor at a time. It is due to the varied revisit time of a satellite. Thus, the images acquired from different sensors temporally are in need to be used for risk assessment models.

Similarly, the satellite images are applied for Land cover change detection applications to monitor the agricultural growth and also to estimate the devastation if any occurred for a long period. It needs highly temporal images for a large scale area which admits huge contrary in image characteristics due to different seasonal conditions but it has been effectively handled using local features like SIFT and SURF [12, 13]. Thus on witnessing the effective role of robust local features in different viewing conditions, the proposed work is aimed at exploiting the use of SURF feature points for effective land cover detection in multi-sensor moderate resolution images. The performance of the SURF descriptors has been evaluated using multi-sensor images with noise and dissimilarities. The three different types of noise such as salt and pepper, speckle and Gaussian are applied for evaluating the performance in noisy images. The additional details and related works on the local features applied in satellite image analysis are illustrated in the following Sect. 2 and the proposed work is illustrated in Sect. 3.

2   Related Work

The robust local features are widely applied in Object Recognition/Detection, 3D. Reconstruction, satellite image registration and image mosaicking applications [14, 15]. The conventional satellite image registration methods suffer with severe limitations on scale and intensity variations [16] and to overcome these limitations, the use of local features is adopted. It is evident that the tedious processing of image registration is highly automated using SURF feature descriptor [17]. The very high resolution satellite images would have more complexities on geometrical properties. It needs automated image registration or object detection schemes which can be achieved through fast and robust local feature extraction and matching methods [18] since the predominant steps of both image registration and object recognition are feature extraction and feature matching.

On image registration, two or more images of same area acquired from different sensors at different time periods with varied geometrical and illumination properties are completely overlaid for better visualization. For example, the fusion of Synthetic Aperture Radar (SAR) and optical images could be a major boon for satellite image analysis since SAR images can be acquired both day and night [19]. Similarly, for image mosaicking the available satellite images in Geotiff format are fused together with region boundaries to obtain mosaic images [20]. The use of robust local features could be a benefit for faster generation of mosaics.

To improve the level of accuracy of objects in case of object detection, suitable image enhancement algorithms like contrast stretching, histogram equalization and color space conversions can be used to achieve a positive effect on feature point detection [21, 22]. The different types of scale restriction and orientation restricted models of SIFT and SURF are also applied in order to improve the performance of local features based object detector [23, 24]. On comparing the performance of SIFT and SURF features in depth maps for hand gesture recognition, it is found that SURF outperforms SIFT [25]. Similarly in a video tracking application, it is found that SURF performs three times faster than SIFT descriptors [26]. Thus, SURF feature descriptors are found to be more efficient than SIFT in different applications. The complexity of satellite image analysis to discriminate individual classes is highly increased in moderate resolution satellite images. For example, many researches are being carried out using non-commercial low resolution open street maps for positioning the vehicles for better safety and navigation [27]. In this paper a similar effort of developing a simple approach for object detection in low resolution satellite images has been proposed using SURF. It is aimed at examining the efficiency robust features for object localization in reduced resolution through which a specified land cover will be detected efficiently in multiple sensor moderate resolutions satellite images.

3   Proposed Work

3.1   Study Area and Data Specifications

The South western coastal region of India that comprises the Pamban island is the chosen study area which is located between 9°11' N and 9°19' N latitude and 79°12' E to 79°23' E longitudes [28]. The images from different sensors like Linear Imaging Self Scanner (LISS III), Advanced Wide Field Sensor (AWiFS) and Enhanced Thematic Mapper + sensor (ETM+) images are used. The specifications of these satellite images are shown in Table 1.

The LISS III and AWiFS images are acquired from the Bhuvan open data portal [29]. The Landsat ETM+ data are acquired from Earth explorer site [30]. In addition to

Table 1. Data specifications

Data	Satellite	Resolution	Bands
LISS III	Resourcesat	23.5 m	4
AWiFS	Resourcesat	56 m	4
ETM+	Landsat 7	30 m	11

satellite images, Normalized Difference Vegetation Index (NDVI) developed from MODIS data acquired from NASA/GSFC, Rapid Response site [31] are also tested.

3.2 Phases of Work Flow

The proposed object detection approach in multiple sensor satellite images hold the following phases: (1) Data Preprocessing, (2) Input Selection, (3) SURF based Feature Extraction, (4) Feature Matching and (5) Object Detection as shown in Fig. 1.

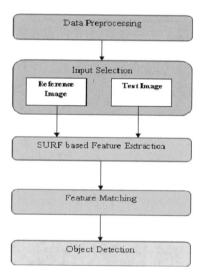

Fig. 1. Proposed approach

3.2.1 Data Preprocessing

The LISS III and AWiFS data hold 4 bands with varied spatial resolution. The combination of bands 3 and 4 which are the Red and Near Infrared (NIR) bands respectively is highly used for vegetation assessment in agricultural applications using the following Eq. (1) in QGIS software.

$$(B4 - B3) / (B4 + B3) \tag{1}$$

Similarly, NDVI tiles derived from MODIS data of the chosen study area with 2 km resolution are combined to generate the Mosaic images.

3.2.2   Input Selection

The reference image and test image are selected from the chosen set of multisensor images to perform land cover detection. A reference image comprises of the specified region to be detected and the test image is the image which contains the reference image.

3.2.3   SURF Based Feature Extraction and Matching

The SURF is a robust scale and rotation invariant feature detector which has been partly inspired by the previous SIFT descriptor [32]. It is based on Hessian & Integral Images. It uses a Binary Large Object (BLOB) detector based on Hessian to find the corresponding Interest points. A Hessian matrix is termed for a square matrix of second order partial derivatives of a function. The determinant of this Hessian Matrix could express the local change around the area. In 2001, Viola Jones adapted an idea of Haar wavelets and developed the Haar-like features. A simple rectangular Haarlike feature is defined to be the difference of sum of pixels of areas inside the rectangle of any position and scale within the original image which is termed for Integral Images [33] as shown in Fig. 2.

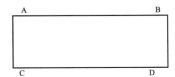

Fig. 2. Integral image

It is computed in recursive manner, to minimize computation and it starts from the top-left and works down row-wise. Thus, image intensities for rectangle of any size in an image can be computed quickly as given in Eq. (2).

$$\text{Total Image Intensity} = I(D) + I(A) - I(B) - I(C) \qquad (2)$$

This SURF features are detected in both the test and reference images individually and stored for matching. In order to locate the object of interest from reference image in the test Image, the reserved SURF features are analyzed for matched pairs. In matching stage, the trace of the Hessian matrix computed at the detection stage for the corresponding interest point is considered. Hence, it achieves faster matching without affecting the descriptor's performance. If the number of matched pairs is found to be very less, then the object detection will fail.

3.2.4   Object Detection

On successful matching of the image pairs, the specified region in reference image is found to be detected in test images within a bounding box which clearly segments the target from the rest of the image. Since SURF is a rotation invariant feature descriptor, the objects can be easily detected in satellite images of varied spatial resolutions. The different combinations of image pairs that are analyzed under variable conditions to evaluate the performance of SURF are listed in the following Sect. 4.

4   Results

The reference and test images are evaluated under four different conditions: (a) Images from Same Sensor, (b) Images from Different Sensors, (c) Images with Noise and (d) Images with Dissimilarities.

4.1   Images from Same Sensor

Both the reference and test images chosen from the same sensor satellite images have been tested. The AWiFS sensor images are analyzed for land cover detection as shown in Fig. 3. Similarly, the results using Landsat images are shown in Fig. 4. The object detection is found to be good in this case using SURF but may be affected by increased computation time for huge dimensional data as shown in Table 2.

Table 2. Object detection results for similar sensor

Reference image	Test image	Object detection	Computation time
AWiFS	AWiFS	Yes	15.19 s
Landsat	Landsat	Yes	33.37 s
NDVI	Mosaic	Yes	9.98 s

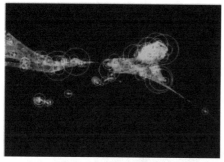

(a) AWiFS Reference Image   (b) AWiFS Test Image

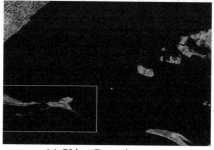

(c) Object Detection

Fig. 3. (a) and (b) – SURF features detection in AWiFS images; (c) Object detection.

4.1.1   AWiFS Sensor

4.1.2   Landsat – ETM+ Sensor

Among the chosen data for evaluation, the Landsat data is found to be huge in size with increased dimensionality and has consumed high computational cost.

4.2   Images from Different Sensors

On applying the robust SURF based object detection approach, the objects of interest can be easily localized in multi-sensor satellite images. The LISS III and AWiFS data are analyzed and found to be successful as shown in the following Fig. 5.

On successive tests using SURF in different sensor satellite images as shown in Table 3, the object detection is found to be good if the no. of matched SURF features.

'x' lies between $12 \leq x \leq 50$. Hence, the maximum no. of matched features is assumed to be 50 and a match ratio is computed. It is termed as the ratio of the actual no. of matched pairs to the maximum no. of matched pairs (i.e. 50) as given in the following Eq. (3).

$$Match\,Ratio = \frac{Actual\,no\,of\,Matched\,Pairs}{Maximum\,no\,of\,matched\,Pairs} \tag{3}$$

The LISS III images cannot be considered as the test image since the spatial resolution is comparatively lower than the other two sensor images. However, the above results shows that object detection in different sensor images using SURF is found to be good.

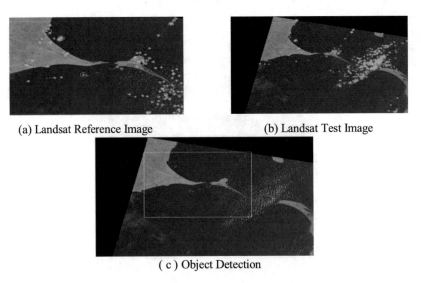

(a) Landsat Reference Image   (b) Landsat Test Image

(c) Object Detection

Fig. 4. (a) and (b) – SURF features detection in Landsat images; (c) Object detection.

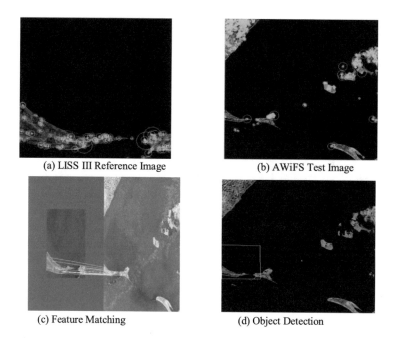

(a) LISS III Reference Image

(b) AWiFS Test Image

(c) Feature Matching

(d) Object Detection

Fig. 5. (a) and (b) – SURF features detection in LISS III image and AWiFS image; (c) and (d) Feature matching and Object detection.

Table 3. Object detection in different sensor images

Reference image	Test image	Object detection	Matched pairs	Match ratio	Computation time
LISS III	AWiFS	Yes	22	0.44	13.6 s
Landsat	AWiFS	Yes	43	0.86	39.2 s
AWiFS	Landsat	Yes	15	0.30	33.5 s
LISS III	Landsat	No	6	0.12	26.5 s

4.3   Images with Noise

The Satellite images are generally affected by high level of noise and illumination conditions. In order to check the performance of the proposed object detection approach, three different types of noise like salt and pepper, gaussian and speckle noise are applied to the reference image. The approach is evaluated by gradually increasing the noise level and found that object detection is highly affected by Gaussian noise as shown in Table 4.

Table 4. Object detection under noise conditions

Reference image	Test image	Noise type	Noise level	Object detection
LISS III	AWiFS	Salt&Pepper	0.04	Yes
			0.06	Yes
			0.12	No
			0.14	No
		Speckle	0.04	Yes
			0.06	Yes
			0.12	Yes
			0.14	No
		Gaussian	Mean = 0; Variance = 0.04	Yes
			Mean = 0; Variance = 0.06	No
			Mean = 0; Variance = 0.12	No
			Mean = 0; Variance = 0.14	No

4.4   Images with Dissimilarities

In addition to noisy conditions, the images with dissimilarities are also considered for evaluating the accuracy of the proposed approach. The Jaccard coefficient is defined to be the ratio of the size of the intersection divided by the size of the union of the chosen images (i.e.) both reference and test images [34]. The similarity ratio for images is computed using Jaccard coefficient given in Eq. (4).

If the similarity ratio is found to be 1, then the images are more similar. The images with very less similarity ratio are chosen and evaluated as shown in Fig. 6 with the corresponding results in Table 5.

$$Jaccard\ Coefficient = \frac{|A \cap B|}{|A \cup B|} \tag{4}$$

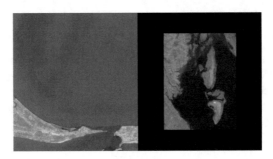

Fig. 6. No feature matching in dissimilar images

Table 5. Object detection in dissimilar images

Reference image	Dissimilar test image	Similarity ratio	Object detection
LISS III	AWiFS	0.1200	No
LISS III	Landsat	0.2200	No
NDVI	Mosaic	0.0700	No

5   Conclusion

In summary, the use of robust local features for increased speed and the need for multi-sensor image analysis for efficient satellite image interpretation has been illustrated. The proposed object detection approach using SURF is found to be efficient as it could overcome the complexities of high computational time and reduced accuracy on interpreting moderate resolution multi-sensor satellite images. To extend this work, more number of satellite images from many different sensors can be applied and comparing the performance of SURF feature descriptor with other feature descriptors.

References

1. Rongqun, Z., Daolin, Z.: Study of land cover classification based on knowledge rules using high-resolution remote sensing images. Expert Syst. Appl. **38**, 3647–3652 (2011)
2. Al-doski, J., Mansor, S.B., Shafri, H.Z.M.: War Impact studies using remote sensing. IOSR J. Appl. Geol. Geophys. (IOSR-JAGG) **1**(2), 11–15 (2013)
3. Widyasamratri, H., Souma, K., Suetsugi, T., Ishidaira, H., Ichikawa, Y., Kobayashi, H., Inagaki, I.: Air temperature estimation from satellite remote sensing to detect the effect of urbanization in Jakarta, Indonesia. J. Emerg. Trends Eng. Appl. Sci. (JETEAS) **4**(6), 800–805 (2013)
4. Baboo, S.S., Shereef, I.K.: An efficient weather forecasting system using artificial neural network. Int. J. Environ. Sci. Dev. **1**(4), 321 (2010)
5. Maqsood, I., Khan, M.R., Abraham, A.: An ensemble of neural networks for weather forecasting. Neural Comput. Appl. **13**, 112–122 (2004)
6. Sarmah, S., Bhattacharyya, D.K.: A grid-density based technique for finding clusters in satellite image. Pattern Recogn. Lett. **33**, 589–604 (2012)
7. Al-Wassai, F.A., Kalyankar, N.V.: Major limitations of satellite images. J. Glob. Res. Comput. Sci., 51–59 (2013)
8. Alhwarin, F., Wang, C., Ristic-Durrant, D., Graser, A.: Improved SIFT-features matching for object recognition. In: BCS International Academic Conference, pp. 179–190 (2008)
9. Bhangale, U.M., Durbha, S.S.: High performance SIFT feature classification of VHR satellite imagery for disaster management. In: IEEE IGARSS, pp. 3574–3577 (2014)
10. van der Werff, H.M.A., van der Meer, F.D.: Shape-based classification of spectrally identical objects. ISPRS J. Photogrammetry Remote Sens. **63**, 251–258 (2008)
11. Li, J., Zhang, Y.: Learning SURF cascade for fast and accurate object detection. IEEE (2013)
12. Flora, D., Delon, J., Gousseau, Y., Michael, J., Tupin, F.: Change detection for high resolution satellite images based on SIFT descriptors and an a contrario approach. MAP5 (2014)
13. Lee, S.R.: A coarse-to-fine approach for remote sensing image registration based on a local method. Int. J. Smart Sens. Intell. Syst. **3**(4), 690–702 (2010)

14. Golash, R., Jain, Y.K.: Dynamic hand localization and tracking using SURF and Kalman algorithm. Int. J. Comput. Appl. **105**(16) (2014)
15. Li, J., Wang, T., Zhang, Y.: Face detection using SURF Cascade. In: IEEE International Conference on Computer Vision Workshops (2011)
16. Hasan, M., Jia, X., Robles-Kelly, A., Zhou, J., Pickering, M.R.: Multi-spectral remote sensing image registration via spatial relationship analysis on SIFT keypoints. In: IEEE IGARSS, pp. 1011–1014 (2010)
17. Bouchiha, R., Besbes, K.: Automatic remote-sensing image registration using SURF. Int. J. Comput. Theory Eng. **5**(1) (2013)
18. Han, Y., Kim, Y., Byun, Y., Choi, J., Han, D., Kim, Y.: Automatic registration of high resolution images in urban areas using local properties of features. In: ASPRS Annual Conference (2011)
19. Fan, B., Huo, C., Pan, C., Kong, Q.: Registration of optical and SAR satellite images by exploring the spatial relationship of the improved SIFT. IEEE Geosci. Remote Sens. Lett. **10** (4), 657–661 (2013)
20. Murali, Y., Mahesh, V.: Image mosaic using speeded up robust feature detection. Int. J. Adv. Res. Electron. Commun. Eng. (IJARECE) **1**(3) (2012)
21. Pramunendar, R.A., Shidik, G.F., Supriyanto, C., Andono, P.N., Hariadi, M.: Auto level color correction for underwater image matching optimization. IJCSNS Int. J. Comput. Sci. Netw. Secur. **13**(1), 18–23 (2013)
22. Vural, M.F., Yardimci, Y., Temizel, A.: Registration of multispectral satellite images with orientation-restricted SIFT. In: IEEE IGARSS, pp. 243–246 (2009)
23. Bastanlar, Y., Temizel, A., Yardimci, Y.: Improved SIFT matching for image pairs with a scale difference. IET Electron. **46**(5), 346–348 (2010)
24. Teke, M., Temizel, A.: Multi-spectral satellite image registration using scale-restricted SURF. In: IEEE International Conference on Pattern Recognition, pp. 2310–2313 (2010)
25. Sykora, P., Kamencay, P., Hudec, R.: Comparison of SIFT and SURF methods for use on hand gesture recognition based on depth map. AASRI Procedia **9**, 19–24 (2014)
26. Anitha, J.J., Deepa, S.M.: Tracking and recognition of objects using SURF descriptor and Harris corner detection. Int. J. Curr. Eng. Technol. **4**(2), 775–778 (2014)
27. Amin, Md.S., Bhuiyan, M.A.S., Reaz, M.B.I., Nasir, S.S.: GPS and map matching based vehicle accident detection system. In: IEEE Student Conference on Research and Development (2013)
28. Study area details. http://en.wikipedia.org/wiki/Pamban_Island
29. Bhuvan Data. http://bhuvan.nrsc.gov.in/data/download/index.php
30. Landsat Data – Earth Explorer. http://earthexplorer.usgs.gov/
31. NDVI Data. http://lance-modis.eosdis.nasa.gov
32. Bay, H., Ess, A., Tuytelaars, T., Van Gool, L.: SURF: speeded up robust features. Comput. Vis. Image Underst. (CVIU) **110**(3), 346359 (2008)
33. Viola, P., Jones, M.: Rapid object detection using a boosted cascade of simple features. In: Conference on Computer Vision and Pattern Recognition (2001)
34. Jaccard Index. http://en.wikipedia.org/wili/ard_index

A Knowledge Extraction Framework for Call Center Analytics

Romeo Mark A. Mateo[(✉)]

Research and Development Group, Nexus Community, 15F Ace Highend Tower
2, Guro 3 dong Guro-gu, 152-724 Seoul, South Korea
romeomarkmateo@gmail.com

Abstract. In the era of Big Data, performing knowledge extraction in large dataset and forming this into readable information such as graphs and story boards are becoming popular for call centers. However, it is not easy to integrate these approaches into the legacies of Computer Telephony Integration (CTI) system which can be used for data analysis. This paper shows a framework that performs knowledge extraction functions to call center data and other data sources such as social networks. The procedures involved are data storing and retrieval, data virtualization and data mining. Before performing knowledge extraction, the needed data are virtualized, and then, the knowledge acquisition module extracts knowledge in form of data patterns where these patterns can be used by call centers for their data reports and analytics. This research paper shows an implementation of components in processing data and knowledge extraction.

1 Introduction

Internet has dominated the global information system in providing vast amount of information but still users need to extract and transform this information into a readable form such as web pages. In spite of the popularity of search engine in finding solution to variety of problems, call assistance is still preferred by serious customers because of the comfort to speak with a person. This is also the reason why business industries continuously having call centers to support their customers to provide personal assistance through conversation. Typically in a call center, Private Branch Exchange or PBX is the hardware behind in automating call switching among agent representatives based on their availability, expertise or other strategies. Primitives like call switching and interfacing are managed in servers or computers by a Computer Telephony Integration or CTI. This is improved by using a CTI middleware to perform complex administration and monitoring of the system using attractive and easy to use GUIs. Additionally, it has a capability to generate beautiful reports and display employees' performance so that it can be used for strategies. While a CTI middleware does the task of administration, other solutions such as social networks [1] and Cloud infrastructure [2] can be used by business establishments to make work easier for employees and to improve collaboration with their business partners.

Social Media have become an excellent tool for businesses by informing customers or consumers about their services, as well as to establish the means of communication

© Springer Nature Switzerland AG 2019
B. K. Ane et al. (Eds.): WSC 2014, AISC 864, pp. 129–141, 2019.
https://doi.org/10.1007/978-3-030-00612-9_12

with their customers [3]. Social Network Services (SNS) can help businesses use sales data, automate marketing decisions and inform consumers of appropriate services based on the data retrieved from the social network sites. There are also efforts in integrating social networks in call center solutions like in [1] because they believe that this will improved the collaboration of agent representatives and customer service [4]. However, this still lack of interoperability in gathering and processing information. It is very difficult and time consuming to read, filter and analyze manually the information that is not from the system's database. Interoperating this process is necessary to include outside information that can be a factor in extracting knowledge.

The objective of this research is to interoperate data processing not only in social networks but also coming from outside sources of a call center for knowledge extraction. This paper presents a knowledge extraction framework for CTI system to store and retrieve data and extract knowledge that is needed in analytics of a call center. The proposed framework gathers information from both CTI system and other data source such as social networks to be stored in data storage that can be a DBMS or NoSQL via accessing Web Application Programming Interfaces (Web APIs). In retrieving and extracting knowledge, these data are virtualized and then data mining is processed. The result from data mining is used by web applications in statistical reports and data analytics.

2 Related Work

2.1 CTI Middleware

CTI middleware solution plays an important role in customizing the needs of a business' call center. The reliability of middleware solution depends on the customer satisfaction in using the products and quality of technical support. Some middleware vendors are offering not only software but also their own hardware products such as phone units to orchestrate their solution. Well known vendors like CISCO [5], Avaya [6] and IBM [7] have been successful in providing variety of solutions according to the needs of their customers and technology trends. Unified Communication (UC) solution integrates all communication products such as chat, mobile messaging and e-mail in a call center. UC is highlighted in the products of CISCO and Avaya. Because of the emergent of Cloud computing, most CTI middleware providers support products in one package where CTI system is installed and configured in virtual machine that eliminates complex installations for ease of deployment. Another solution is a web based system where call centers do not need to install or configure software but only use web browser as their interface to communicate with their customers.

Social networks are also have been used to enhance the collaboration in call centers. An example of using this approach is the Chatter Service Cloud [2] of SalesForce in which it manages call centers and integrates social networks in their interface to provide a quick way to collaborate. However, it does not support data mining integration and lacks of processing semantic information.

Nexus Community is an IT company that has been providing CTI middleware solution to call centers in South Korea since 2001. The flagship product named

Nexus CUBE [8], which is a client/server program, is already expanded in Cloud solutions. The company's products have already reached not only Asia but also Europe markets. Currently, a web-based project is under development in which the latest web technologies are being researched to integrate in its CTI core technology. A knowledge extraction framework is an extension of the system to process CTI data and social network in knowledge extraction. The results from knowledge extraction are used for data reports and other future applications.

2.2 Data Mining in Call Centers

Data mining is usually used in market analyses, fraud detection, weather forecasting and other applications that process large amount of data to extract important information. Call centers used data mining to improve their management system and customer relationship shown in the previous studies. In [9] shows a study of applying data mining approach in determining the performance of call centers. They tested different models of data mining for evaluation and then proposed a hybrid algorithm based on decision tree and neural networks. In improving the interaction between the contact center representatives and clients, information from text or speech contents are subject for analysis to classify knowledge. In [13], they propose a context-aware information retrieval system that facilitates reduction in call resolution time in a contact center. Also, studies show that the classification mining in call centers is improved by adding or integrating meaningful contents [10–12]. In [14] result shows that using basic word contents in classification has limited capability and suggesting further sophistication such as natural language rule may be required. A pragmatic analysis is used in [15] to call center conversations in order to provide useful insights for enhancing data analytics. The method in [15] can detect situations as controversial topics, customer-oriented behaviors and predict customer ratings which are hardly detectable by standard approaches. As data are being accumulated in database storages and Internet is expanding, the methods of retrieving and managing data are also getting more sophisticated. As suggested in [16], using the technologies in Big Data could improve the contact center analytics. In our efforts to prepare call center for Big Data and improving the data reports of CTI System in NEXUSCUBE, data mining module is integrated for analytics of CTI System of NEXUS COMMUNITY.

3 Knowledge Extraction Framework

3.1 Web CTI System

The Web CTI system of Nexus Community is based on a components-based approach in integrating web applications and services. This is abstracted into three layers which are (1) Component Layer, (2) Core Layer and, (3) Data Layer. In the Component Layer, a web application represents a component that is accessed directly by clients which are shown at the top layer in Fig. 1. Other programs or scripts such as web plugins that are servicing the web application are also represented as components. Web applications can also interact with other web applications. These components use a post method to request or process data and a polling engine for event notification in the web servers. In the case of our system, both methods use Ajax calls. Popular frameworks like Spring [17] and SockeIO [18] uses the same technology to implement push notification method with web clients. In the Core Layer, CTI functions are implemented in the web servers. Events from the CTI system are processed in Web CTI and these events become real-time data which will be pushed to clients receiving the event. E.g., calling events will be received by the caller and callee clients that are currently connected to the Web CTI System. If clients are not present then the real-time data for clients are stored in the web server's memory and these can be retrieved later by the polling engine. Data for displaying information and reports are requested in Web Database. Interaction between Web CTI and Web Database are implemented through message passing.

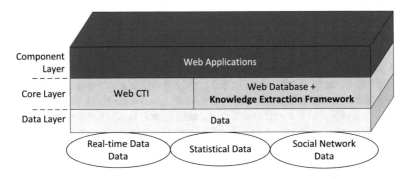

Fig. 1. Web-based CTI System using Knowledge Extraction Framework.

The Web Database extends its function by integrating the Knowledge Extraction framework (KEF). The data from various sources such as CTI database and social networks are processed in KEF to be used for data analytics. Examples of input data from social networks are user profile and post message from a social network sites. The KEF extracts this information based on the requirements of an application that provides data analytics. In Data layer, real time data, statistical data and social network data are illustrated. Real-time data are CTI events which are pushed to clients as real-time interaction with the system while statistical data which include Social Network Data are not used in real-time events.

3.2   Input and Output of Knowledge Extraction Framework

The Data layer shown at the bottom layer of Fig. 1 is using both Relational Database Management System (RDBMS) and Not-only-SQL (NoSQL) storage types. Both RDBMS and NoSQL storages are supported in the proposed framework via libraries. In the Web CTI system, statistical data are handled in RDBMS while data gathered from outside sources such as social networks are stored in NoSQL. The proposed system uses the following policies in storing data from both storage systems:

- RDBMS only contains data from the Web CTI system.
- NoSQL contains data other than the data from the Web CTI System.
- NoSQL can associate data from Web CTI system into data stored in NoSQL storage, e.g., agent information in Web CTI System is associated to users of the Social Networks.

Because of the legacy features of RDBMS, it is still more preferred by application developers and optimal in retrieving complex relations of data. However, NoSQL is ideal in storing very large data considering scalability. The data consistency of user and transactions in Web CTI system are kept in RDBMS while NoSQL exploits the redundant and scalability of storing data. Figure 2 shows data storing using RDBMS API and NoSQL API in the Web Database server.

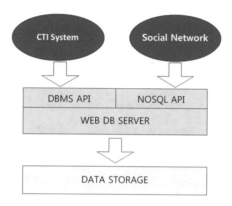

Fig. 2. Storing data using web APIs in the Web Database Server.

The components in CTI system are using web APIs to interact with the data storage. The RDBMS API is used to execute SQL and other sources such as Social Networks Site are using NoSQL API to store and retrieve data in the NoSQL storage. These APIs are integrated in the Web Database Server where it calls the RDBMS and NoSQL libraries to interact with the data storage. Figure 3 shows the process of using KEF.

All data gathered in the Web CTI System and outside sources are stored in data storages. In Fig. 3, the data are processed through retrieving data by a data extractor, virtualizing data in the form of objects and then extracting knowledge in the virtualized data. In virtualizing data, tables are replicated in the form of virtual tables before it can be requested for knowledge extraction. The virtual tables can handle SQL queries and updates from client request. For data analytics, the knowledge extractor is called upon request to perform data mining in each data request. An example of message packet is presented in Fig. 4.

In Fig. 4, the inputs of request message packet contains *object, execute* and *command*. The *object* addresses the component in the web server and *execute* specifies the method of the component. The command specifies the SQL table which selects the data that will be performed and a sample message packet output is shown in Fig. 5.

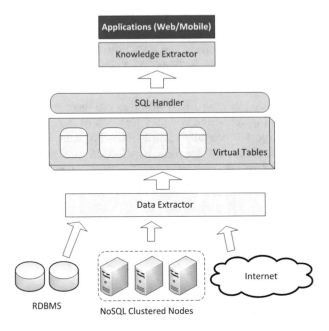

Fig. 3. Retrieving processed data using Knowledge Extraction Framework.

```
<?xml version='1.0' encoding='utf-8'?>
<rio>
        <invoke id='1' object='extractKnowledge' execute='pattern'>
                <set id='command'>'SQL Statement'</set>
        </invoke>
</rio>
```

Fig. 4. Message format in requesting KEF

The response message packet contains data rows which consist of columns. *premise* and *consequence* are outputs from request in Fig. 5 to represent extracted association rule. The data mining methods in KEF are discussed in the next section.

3.3   Analytics Using Data Mining

Data mining method has proven to be useful in extracting data patterns and classifying information to predict market trends and detect data anomalies in companies. An early study of integrating data mining in call center is used to analyze the performance of a call center in [10]. Current features are not only focused on analyzing agent's performance but also other media such as speech recognition, text analysis and data analytics. This only shows that data mining is gaining popularity in enhancing the

```
<?xml version='1.0' encoding='utf-8'?>

<rio>

        <notify queryName='pattern'>

                <row id='1'>

                        <set id='premise'>

                        itcall_rcount EQUALS 0-22.8 AND grptrcall_time EQUALS 0

                        </set>

                        <set id='consequence'>grptrcall_rcount EQUALS 0</set>

                </row>

        </notify>

</rio>
```

Fig. 5. Message format of a response from KEF.

feature of a CTI provider in able to compete with other CTI providers. In the framework, common algorithms are integrated in the data mining module of the Knowledge Extraction Framework.

Clustering and Classification. Clustering method is a data mining method that groups the data with similar properties based on some constraint. The procedure analyses the properties of data and groups the data that have similar properties based on a specified constraint. k-means clustering is a common clustering technique that simply uses k value which is the number of groups to be constructed and then groups each n data to the nearest group by Euclidean distance function. Given a set $X = \{x_1, x_2, \ldots, x_n\}$ where each data is a d-dimensional real vector, k-means clustering aims to group X into k ($\leq X$) sets which are $g = \{G_1, G_2, \ldots, G_k\}$ so as to minimize the within-cluster sum of squares. The goal of the function shown in Eq. 1 is to minimize the mean squared error (MSE) in each group.

$$func(X) = \min \sum_{i=1}^{k} \sum_{x_j \in G_k}^{X} |x_j - \mu_i| \qquad (1)$$

In Eq. 1, the μ_i is which the mean value in group G_k is subtracted to x_j and then calculate the total value in each group. The lowest value from this function will be the selected grouping of data. The structure from the group is used to classify each data pattern to identify where a data belongs. In this research, the output from clustering are used to group the agent performance by calling the data of agent performance and perform clustering on it.

Association Rule Mining. Association mining is another data mining technique that extracts frequent patterns from data. Association rule mining is used to determine the frequent patterns. These frequent itemsets in the transactions of accessed services are extracted using the Apriori algorithm. The algorithm cannot process numerical type of data and so a categorize method is used in each attribute columns. First, the minimum value and maximum value are looked in the current records in the column. Then, the range is calculated by subtracting minimum value to maximum value and divided by the number of desired category. Equation 2 is used in determining the range value to categorize the numerical values.

$$r(A) = \frac{\max_A - \min_A}{\omega} \tag{2}$$

In Eq. 2, A are distinct values from all data in a column. r is a function to process column values in A in which the highest value max_A and lowest value min_A are determined in the function. min_A is subtracted to max_A to get the value of the range and then this is divided into a number of categories specified by ω. The next step to this is calculating the Apriori algorithm. There are two steps of the Apriori algorithm which are; (1) join step which generates the candidate patterns by providing all the combinations from each itemset and (2) prune step which chooses the frequent pattern from the combination sets. The join step finds L_k, a set of candidate k-itemsets by joining L_{k-1} with itself. The prune step generates C_k as superset of L_k, and all of the frequent k-itemsets are included in C_k. Choosing the frequent patterns are based on support count and selecting the rules uses a confidence value. The support count (s_count) determines the frequency of each pattern shown in Eq. 3 where A and B are two items from L_k. The probability of AUB is shown in Eq. 3. After the support count, the confidence is determined by getting the ratio of support count of each item (A) in an itemset (AUB) shown in Eq. 4. The A->B is a rule extracted after calculating the confidence.

$$s\_count(A, B) = P(A \cup B) \tag{3}$$

$$conf(A \rightarrow B) = \frac{s\_count(A \cup B)}{s\_count(A)} \tag{4}$$

After pruning, the rules are collected in R to be outputted in the response message. In our proposed framework, these algorithms are called in data mining module for used by any application in the application layer by including the method in the request message. Based on the request call, data from both CTI and social network sites are processed using data mining techniques. E.g., adding the name of algorithm in the invoke method attribute will include processing the request data into the specified algorithm, execute=pattern calls the Apriori Algorithm. In this research, the output from association is intended for predicting the call patterns of agents.

4   Implementations

The Web CTI System was already included in our product named NEXUS CaiRO (All-In-One solution). In addition, the Knowledge Extraction Framework (KEF) was integrated in the Web CTI system to support the knowledge extraction in CTI data. Both Web CTI system and KEF were built using Java programming and used the servlet specifications to implement in Apache Tomcat Server.

Figure 6 shows the main menu of the Web CTI system which is used to link several web applications. Before displaying the main menu, an authentication procedure is required by entering the agent identification and password. Currently, two web applications which are *Administrator* and *Report* are being used in our product. The *Administrator* is a basic management for agents, groups, tenants, agent skill information and media configuration. These are viewed in grids/tables provided with add, edit, delete and sorting functions. *Report* is another application which is used for displaying statistical reports of a call center. In the *Report*, complex SQL statements are used to request the data and displayed it in texts or graphs.

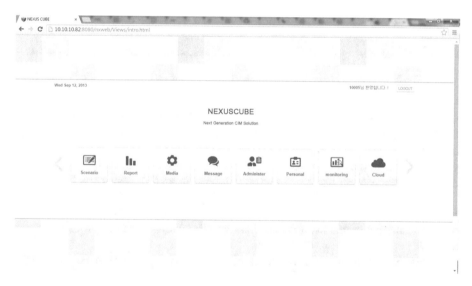

Fig. 6. User interface of Web CTI system showing available applications.

A web API for social network was also integrated in the KEF to store data from social network pages. Figure 7 shows the blog page which is used by agent representatives to read and track issues about their clients and peer agents. This blog page is using the web API in KEF to manage users and post messages and it stores data in a NoSQL storage type. An administration page is used to register agents in able to read and post on the blog.

A web application using the KEF is shown in Fig. 8. The KEF is used to group or classify agents based on their performance. As explained in Sect. 3.2, an application

Fig. 7. A blog page for handling technical issues which is accessed by call center agents.

uses POST method to access the interface of KEF. The total call counts and time spent in a call were used for the agent performance. A table, named *agent_calls*, from the main database was used which contains 68 columns with 21,861 data rows. This table contains call activities of agents such as inbound and outbound calls. Each row is a single transaction of an agent and these are totaled in an SQL using the GROUP BY function. The *agent_calls* will be replicated in the virtual tables and then, after totaling, these are processed in data mining. The clustering function and EM algorithm was used to dynamically determine the number of groups. The output from this process is shown in Fig. 8.

In Fig. 8, the *agent_calls* is processed in data mining where agent is grouped according to their similarities. Each line bar in a graph in Fig. 8 is the mean value of each group to represent the groupings. In processing clustering, three groups were identified which is used to classify the data from *agent_calls*. Each group varies in call performance, e.g., inbound calls (IBCALL) in *Data Group 1* is lesser than *Data Group 2* and *Data Group 3* but outbound calls (OBCALL) is lesser in *Data Group 3* than *Data Group 1* and *Data Group 2*. In Fig. 9, classification and association mining is applied in the data. On the top of Fig. 9, the statistics of an agent and its group labeled are shown. The lists of agents shown in the middle part of Fig. 9 are grouped based on clustering structure in Fig. 8. Clicking an agent in the list will display the statistic of that agent shown on the upper part of Fig. 9. Lastly, the association rules are extracted on the bottom part of Fig. 9. These association rules can be used to predict the performance patterns of agents.

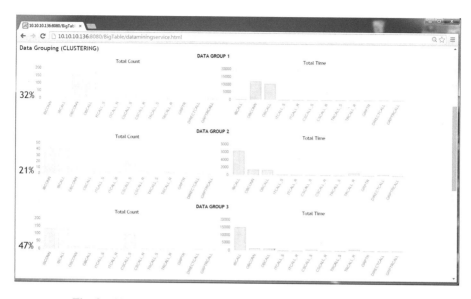

Fig. 8. Clustering result using the total count and total time of calls

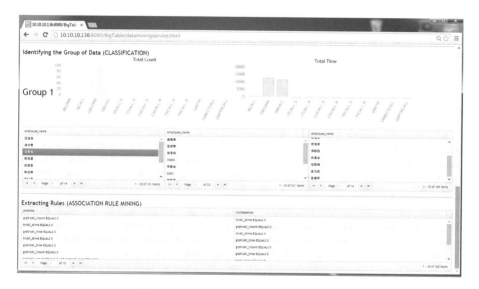

Fig. 9. KEF is used in classification and association mining.

The process in data virtualization is a critical process in KEF before performing data mining or analytics. The response time performance of this process was evaluated by simulating client queries performing select statement in virtual tables. In the performance test, the same table *agent_calls* was used. Three technologies were compared which are Java Database Connectivity calls (*JDBC*), *MyBatis* and virtual tables (*nxTables*). A REST client was prepared and the Web Database Servlets were integrated with libraries of *JDBC*, *MyBatis* and *nxTables*. The query time performance of each method was recorded which is calculated by the total time of the client started the query and client received the response. The first case used a single client performing 300 queries while the second case tested the concurrent request from ten clients performing 300 queries. Figure 10 shows the result from two cases.

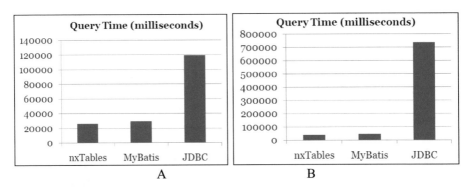

Fig. 10. Performance results in response time using single client (A) and 10 clients (B).

Results in both Fig. 10a and b show that JDBC has the slowest response time and consider not efficient. Both *nxTables* and *MyBatis* can optimize the performance in data retrieval but *nxTable* has slightly improvements in the response time.

5   Conclusion

In this paper, a framework for extracting knowledge is shown for call center's data where these knowledge or extracted data patterns can be used by call centers for their data reports and analytics. The framework supports storing data in the CTI system and outside sources. Data are virtualized represented by objects for fast response in applying data mining. The framework supports the usage of various data mining algorithms called in the message packets to be used by application in providing analytics to call centers. Data virtualization was evaluated by comparing to common DB calls like JDBC and other data virtualization like MyBatis by getting the response time performance.

This research paper shows an implementation of components in processing data and knowledge extraction. The NXWCUBE, name of the Web CTI system, was already included in their Cloud solution products. However, the proposed framework is still a subject for testing for later integration our products.

References

1. Salesforce Chatter. http://www.salesforce.com/eu/chatter/overview/
2. Amazon Web Services EC2. http://aws.amazon.com/ec2
3. Mohr, C.: I'll Connect You: Social Networks Improve Call Center Productivity and Customer Relations. http://call-center-services.tmcnet.com/topics/call-center-services/articles/363201-ill-connect-social-networks-improve-call-center-productivity.htm
4. Ancier, B.: 3 Brands that Use Social Media in Customer Service to Boost Happiness. http://www.tintup.com/blog/3-brands-that-use-social-media-in-customer-service-to-boost-happiness/
5. Cisco Unified Contact Center Enterprise. http://www.cisco.com/c/en/us/products/customer-collaboration/unified-contact-center-enterprise/index.html
6. Avaya Aura Platform. http://www.avaya.com/usa/product/avaya-aura-platform
7. IBM Call Center for Commerce. http://www-03.ibm.com/software/products/en/Call-Center
8. NEXUSCUBE. http://support.nexus.co.kr/en/nexus/nexuscube/nexuscube.php
9. Paprzycki, M., Abraham, A., Guo, R., Mukkamala, S.: Data mining approach for analyzing call center performance. In: 17th International Conference on Innovations in Applied Artificial Intelligence, pp. 1092–1101 (2004)
10. Tang, M., Pellom, B., Hacioghu, K.: Call-type classification and unsupervised training for the call center domain. In: Automatic Speech Recognition and Understanding (2003)
11. Tan, P.N., Blau, H., Harp, S., Goldman, R.: Textual data mining of service center call records. In: 6th ACM SIGKDD International Conference on Knowledge Discovery and Data Mining, pp. 417–423 (2000)
12. Busemann, S., Schmeler, S., Arens, R.G.: Message classification in the call center. In: 6th Conference on Applied Natural Language Processing, pp. 158–165 (2000)
13. Thawani, A., Gopalan, S., Sridhar, V.: Web-based context aware information retrieval in contact centers. In: Web Intelligence, pp. 473–476 (2004)
14. White, S., Jagielska, I.: Investigation into the application of data mining techniques to classification of call centre data. In: The 2004 IFIP International Conference on Decision Support Systems, pp. 793–802 (2004)
15. Pallotta, V., Delmonte, R.: Interaction mining: the new frontier of customer interaction analytics. In: New Challenges in Distributed Information Filtering and Retrieval Studies in Computational Intelligence, vol. 439, pp. 91–111. Springer, Heidelberg (2013)
16. Masterson, M.: Cracking the Big Data Nut in the Contact Center. http://www.smartcustomerservice.com/Articles/News-Features/Cracking-the-Big-Data-Nut-in-the-Contact-Center-97325.aspx
17. Spring Framework. http://projects.spring.io/spring-framework
18. Socket IO. http://socket.io/

Theoretical Development of Evolutionary Algorithms

Soft LMI-Based H∞ Control with Time Delay

Farnaz Sabahi[1,2(✉)] and M.-R. Akbarzadeh-T.[1,2]

[1] Department of Electrical and Computer, Urmia University, Urmia, Iran
f.sabahi@urmia.ac.ir, akbazar@um.ac.ir
[2] Center of Excellence on Soft Computing and Intelligent Information
Processing, Department of Electrical Engineering,
Ferdowsi University of Mashhad, Mashhad, Iran

Abstract. One of the main problems underlying much optimization theory is local optimum. When major parametric uncertainties such as time delays, which are frequently encountered in physical systems, are presented, this problem becomes much more serious. In such situations, evolutionary optimization algorithms may be used as an attempt to overcome this difficulty. In this paper, genetic algorithm (GA) has been adopted to tackle the stated problem in the framework of robust H∞ control. GA is employed to find suitable feedback gains and delay-dependent linear matrix inequality (LMI) solvers to resolve issues related to stability conditions. In addition, to balance between exploration and exploitation, particle swarm optimization (PSO) and ant colony optimization (ACO) first individually and then as a hybrid are augmented with GA. Performance of these hybrid approach is then made with results obtained by only GA approach. The evolutionary LMI-based H∞ control scheme is applied to a single link robot arm. The controller satisfies the desired properties for not only any unknown-but-bounded disturbances but also any uncertain-but-known constant bounded time delay.

1 Introduction

The idea of H∞ optimization in control system goes back to the early days of robust control theory, at least as early as 1979 [1]. The underlying objective of robust control is considering plant uncertainty explicitly to achieve specified level of performance in which a trade-off is sought between robustness and optimality. Some of the related works in this area consider a range of operating points by regarding a variation of uncertainty structure. *M*-synthesis and LMI-based H∞ control are examples of this approach. The former is suitable when uncertainty has a block diagonal structure of uncertain variables and the latter maps the operating points into the uncertainty space and uses this map in the controller to minimize the upper bound of robust performance index. Indeed, LMI-based methods afford a powerful formation for a wide variety of control issues including the H∞, the mixed H2/H∞, the gain scheduling, and many others. But the success of all these methods, in general, and H∞ control, in particular, hinges on characterizing the global minimum of the index norm as well as in managing the problem of stability.

In fact, considering the bilinear character of some constraints in the optimization problem leads to the idea of incorporating evolutionary algorithms, which the familiar

© Springer Nature Switzerland AG 2019
B. K. Ane et al. (Eds.): WSC 2014, AISC 864, pp. 145–157, 2019.
https://doi.org/10.1007/978-3-030-00612-9_13

one is genetic algorithms (GA) [2]. Among works that use this approach in robust control domain is [3] where GA attains the control parameters of the PI controller subject to H∞ constraints in terms of linear matrix inequalities. In [4], fitness function of GA is the results of the linear matrix inequality solver in the framework of H2/H∞ robust control. Another method [5] employs genetic algorithms to linearize the non-linear channel of the FIR equalizer and uses linear matrix inequality to resolve the approximate error of the linear model. Moreover, in reference [6], to solve the H∞ control problem, a method by introducing output feedback in continuous-time linear systems subject to parameter uncertainties and time domain constraints is proposed. In [7], an approach is described that uses the matrix rank minimization problem and performance index as optimization objective by combining genetic algorithm and linear matrix inequality in low order suboptimal H∞ control. Another attempt can be found in [8], where hybrid approach of GA and LMI is used to solve non-convex optimization in mixed H2/H∞ control domain.

The next concern that can be important in control design is the *time delay* problem. Although time delay may be short, it limits the controller performance and may lead to system instability. In [9], a robust genetic control design is proposed that applies a class of uncertain neutral delay system by proposing a delay-dependent stabilization criterion using Lyapunov function, and in [10], a H∞ based controller considering time delay in the system's input is described. In reference [11], the H∞ performance for building structures is shown to be realizable by a delay-dependent feedback control, using hybrid of GA and LMI. These methods report satisfactory results in the presence of time delay.

Another point that should be considered is the deficiencies of GA. While using genetic algorithms, by searching towards the global solution, can handle the problem of local optimum, it still suffers from several deficiencies such as slow rate of convergence and premature convergence. Dealing with these shortcomings is an interesting field of research. Some recent works have proposed where GA is executed for several times, but more interesting techniques aim to find proper tuned parameters of GA during any attempt of search [12]. In [13], the indirect shared memory in ant strategies is borrowed to explore the smallest population of genetic algorithms. Reference [14] explains the hybrid of genetic algorithms and ant colony optimization (ACO) [15] to enhance the performance of just-in-time optimization of concrete-readymade delivery, where ACO solves truck assignment problem. Another hybrid approach to improve the performance of GA is proposed in [16], where particle swarm optimization(PSO) [17] is applied to enhance GA's individuals; this combination is also called GSO.

Here, a hybrid method is described that consists of a genetic algorithm and linear matrix inequality in the H∞ control framework, to explore the bilinear relation between the controller gain and Lyapunov matrix. In this method, the stability conditions in the presence of input time delay are considered by linear matrix inequalities. The performance index is regarded as optimization objective and the search for output feedback gain is done by genetic algorithms. Additionally, three hybrid global-local strategies: GA with ACO, GA with PSO, and GA with ACO-PSO are examined to handle the aforementioned shortcomings. Although the proposed hybrid algorithm involves several additional equations, they are very easy to implement and produce little more computation. In other words, the consumed time for running the algorithms changes little due to the parallel nature of evolutionary algorithms.

The arrangement of this paper after this introduction is as follows: the second section is dedicated to preliminary assumptions about the system and the formulation of the H∞ control problem based on LMI. The third section studies a method that computes a control law based on GA search and implicitly considers stability conditions in the presence of time delay with taking advantage of LMIs. The hybrid algorithm of GA with PSO, ACO, and PSO-ACO for the proposed LMI-based H∞ control scheme is explained in Sect. 4. Section 5 illustrates simulation results of the proposed frameworks. Finally, conclusion is in Sect. 6.

2 Problem Statement

Let us consider a linear time-invariant plant $P(s)$ that maps inputs to outputs, and is described by [18]:

$$\begin{bmatrix} z(s) \\ y(s) \end{bmatrix} = P(s) \begin{bmatrix} w(s) \\ u(s) \end{bmatrix} \tag{1}$$

where w is exogenous input, u is control input, z is controlled output, and y is measured output subject to physical constraints. The suboptimal H∞ control problem is to find output feedback control law, $u = Ky$, such that for a given scalar $\gamma > 0$ the closed-loop transfer function from w to z would be strictly less than γ.

$$\|T_{zw}\|_\infty < y \tag{2}$$

where $\| \ \|\infty$ denotes the H_∞ norm. The general framework of a plant and its controller is shown in Fig. 1. Δ represents uncertainty.

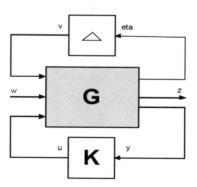

Fig. 1. Uncertainty structure

Let us consider a minimal realization of the plant $P(s)$ as [18]:

$$P(s) = \begin{bmatrix} D_{11} & D_{12} \\ D_{21} & D_{22} \end{bmatrix} + \begin{bmatrix} C_1 \\ C_2 \end{bmatrix}(SI - A)^{-1}(B_1, B_2) \tag{3}$$

where the proper dimensions are $A \in R^{n \times n}$, $D11 \in R^{p1 \times m1}$ and $D_{22} \in R^{p2 \times m2}$. The following assumptions are considered:
(1): (A, B_2) is stablizable and (A, C_2) is detectable, and (2): $D_{22} = 0$.
With any controller $K(s)$ in the following form:

$$K(s) = D_K + C_K(SI - A_K)^{-1}B_k \tag{4}$$

The closed transfer function from w to z is obtained by:

$$T_{zw} = D_{cl} + C_{cl}(SI - A_{cl})^{-1}B_{cl}$$

$$A_{cl} = \begin{bmatrix} A + B_2 D_k C_2 & B_2 C_k \\ B_2 C_k & A_k \end{bmatrix}, \quad B_{cl} = \begin{bmatrix} B_1 + B_2 D_k D_{21} \\ B_k D_{21} \end{bmatrix}, \tag{5}$$

$$C_{cl} = [C_1 + D_{12} D_k C_2 \quad D_{12} C_k], \quad D_{cl} = D_{11} + D_{12} D_k D_{21}$$

By using the bounded real lemma (BRL), the H_∞ norm constraint (2) and the internal stability can be induced from the feasibility of the following matrix inequality:

$$\begin{bmatrix} A_{cl}^T P + P A_{cl} & P B_{cl} & C_{cl}^T \\ * & -\gamma I & D_{cl}^T \\ * & * & -\gamma I \end{bmatrix} < 0 \tag{6}$$

where $P > 0$ is symmetric matrix and the notation $*$ indicates terms that can be induced by symmetry. If the parameters in A_{cl}, B_{cl}, C_{cl}, and D_{cl} are known then the above matrix inequality (6) is LMI.

Now, let us consider state space description of time-invariant system involving input time delay as:

$$\begin{aligned} \dot{x} &= Ax + Ew + Bu(t - \tau) \\ z &= C_z x + D_{zu} u(t - \tau) \\ y &= C_y x + D_{yw} w + D_{yu} U(t - \tau) \end{aligned} \tag{7}$$

where x is referred as the system state. The following theorem provides sufficient conditions for existence of delay-dependent output feedback of H_∞ control in LMI form, with considering controller as u = $Ky = KCx$.

Theorem [11]: *Given a scalar τ_1 the closed-loop system is asymptotically stable with H∞ performance index, > 0, for any constant time-delay satisfying $0 < \tau < \tau_1$, if there exist matrices K, $P > 0$, $Q > 0$, $Z > 0$, X and Y satisfying the matrix inequalities*:

$$P > 0, \quad \begin{bmatrix} X & Y \\ * & Z \end{bmatrix} \geq 0,$$

$$\begin{bmatrix} A_{cl}^T P + P A_{cl} + \tau_1 X + Y + Y^T + Q & PBKC - Y & PE & \tau_1 A^T Z & C_1^T \\ * & -Q & 0 & \tau_1 (KC)^T B_1^T Z & (KC)^T D_{12}^T \\ ** & -\gamma^2 I & \tau_1 B_1^T Z & 0 \\ ** & * & -\tau_1 Z & 0 \\ ** & ** & -I \end{bmatrix} \quad (8)$$

The theorem states that the H∞ performance, regardless of the unknown constant delay, is guaranteed within feasible maximum bound delay, and it is possible to design a delay-dependent feedback controller based on LMI [11].

The stability conditions of theorem have the benefits of not requiring tuning parameters in the case of the LMI solutions. If the output feedback gain K is given, then the problem is said to be convex.

3   GA Formulation of H∞ Control

3.1   Convex Formulation

When feedback gain K is not given, matrix inequalities (8) are bilinear matrix inequalities (BMIs), and so the problem is not convex. GA is an optimization method that can cover a large area of non-convex problems with its most worthwhile aspects of requiring just a cost function and a set of variables, *i.e.* it needs little prior knowledge of the problem in order to find suitable solutions. However, many decision variables increase computation in GA. In other words, GA cannot give solution for K and P at the same time for (8). However, if we know K, (8) is convex and so LMI solver can find P for (8). Therefore, we use GA to seek K and then use LMI solver to solve P to guarantee stability.

3.2   GA Algorithm

The continuous (real-coded) GA has several advantages such as large search space and high precision [19]; therefore, this type is used here. The diagram of the algorithm is shown in Fig. 2. The steps of the applied GA algorithms are:

Step 1:   *Encoding*: Each feedback gain vector K is considered as an individual. There is restriction on the range of gain.

Step 2:   *Initialization*: The N individuals, Ki's, are generated randomly.

Step 3:   *Fitness assignment and Evaluation*: The fitness is defined as inverse of minimum t where t is computed by *feasp* function. The function *feasp* solves the following auxiliary convex program [20]: Minimize t subject to $L(x) - R$ $(x) < tI$, where x is the vector of (scalar) decision variables.

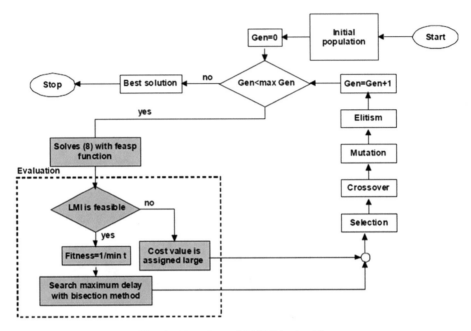

Fig. 2. Flowchart of LMI-GA algorithm

Evaluation process is a little different and divided in two cases.

Case I: When K_i is infeasible, *i.e.* it *does not* fulfill (8): It is not necessary to determine admissible delay. In this case, the cost value that corresponds to it is assigned large enough in order to discard this feedback gain.

Case II: When K_i is feasible, *i.e.* it *does* fulfill (8): The maximum admissible time delay, τ_1, would be computed by the bisection method [11]. The fitness of K_i is determined according to the cost value. Minimizing the cost function is equivalent to reaching a maximum fitness in the search space. This step is repeated for the number of individuals.

Step 4:   *Selection:* To produce new offspring for the next generation, two chromosomes with a selection method are chosen to be involved as parents. The selection is executed based on roulette wheel method.

Step 5:   *Crossover:* Selected individuals are recombined, through linear heuristic crossover method to guarantee that (8) is still satisfied. This method produces a single

$$K_k = \beta(K_{j+1} - K_j) + K_j \qquad (9)$$

where β is a random number in the range [0,1]. In this method, any variable that is outside of the bounds is not considered in favor of the other two. The best new offsprings are propagated.

Step 6:   *Mutation*: To make sure that GA searches the solution space freely, uniform mutation is used with the following rule:

$$K_k = \beta_j \Delta P + K_j \qquad (10)$$

ΔP denotes the increase of individuals and βj is chosen from interval $[-1,1]$. The mutated gene from the search interval is drawn randomly.

Step 7:   *Elitism*: The best individuals in the population are retained in the next generation.

Steps 3–7 go on until the number of generation is reached.

4 Cooperation GA-LMI with Other Population-Based Algorithms

Exploration (diversity preservation) and exploitation (convergence) are two competing goals of global search techniques. In the absence of a prior knowledge, information about the problem is collected by making measurements at a number of points that are distributed over the solution domain uniformly. In other words, exploration is important to make sure that searching is sufficient to find a reliable estimate of the global optimum. On the other hand, the importance of exploitation is to produce better solution by searching the neighborhoods of the best solution. To strike a balance between these two goals, *i.e.* exploration and exploitation, global search methods are combined with other search algorithms and establish global-local search. These hybrid algorithms seem simple enough to implement, yet complex enough to give confidence to improve efficiency and accuracy of the search process.

It should be noted that, GA might cause degeneracies in search performance if their parameters are not carefully chosen. In [13], it was shown that choosing improper fixed size population causes premature convergence. Therefore, the hybrid algorithm that adds ACO to GA in the beginning of the algorithm for choosing its major affecting factor, *i.e.* population size, is proposed here. In this approach, ACO estimates the minimum population size. In fact, the proper population size of GA is gained by the global best solution of ants in ACO algorithm. This ant algorithm is often used to solve Traveling Salesman Problem (TSP) problem kind. Here, every city is a representation of a population size, and the best tour is one that yields the best population size, i.e. one that produces best solution with least computations. Due to page limitations, the flowchart of GA-ACO algorithm is omitted here.

Furthermore, the combination of GA and PSO leads to a suitable hybrid method of cooperation of both. One of them can be used as a pre-optimizer for another. In the hybrid algorithm used here, certain percent of the population in the next generation is occupied by the PSO and the remained population is considered by GA crossover operator (*i.e.* GSO) [21]. This percentage of population is expressed by a driving parameter denoted by kp. In each generation, the fitness of all individuals is calculated; then after dividing population into two sections, each section is evolved by one of GA or PSO algorithms. The two parts in the resulting new population recombined and then in the next generation divided into another two groups randomly, then again each group moves on the solution space by PSO or GA. Tournament selection scheme is adopted here and crossover strategy is chosen linear crossover. Figure 3 shows the flow diagram of the described algorithm.

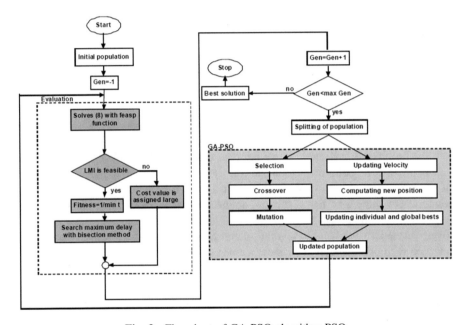

Fig. 3. Flowchart of GA-PSO algorithm PSO

Now, we take advantage of combining two described methods. That approach offers better balance between the exploitation of the search experience gathered so far and the exploration of unvisited search space area. Cooperation between two algorithms goes on at two levels: modification and evaluation. In fact, in the proposed approach, the main improvement is due in first to the efficient use of ACO by adjusting population size and due in second to the procedure of updating next generation by decomposition of population and then taking advantage of both PSO and GA. Figure 4 shows the flowchart of the described algorithm.

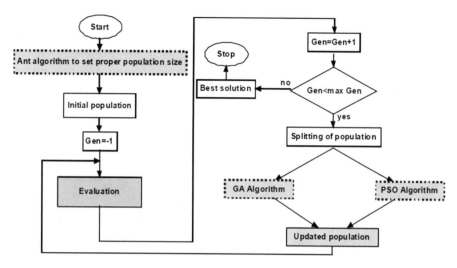

Fig. 4. Flowchart of GA-ACO-PSO algorithm

5   Numerical Results

The performance of the methods is investigated for a one-link robot arm with the following second order differential equation computed by Newton's second law:

$$m\ddot{x} + c\dot{x} + kx = u, m = 5(1 + \Delta_m), k = 5(1 + \Delta_k) \tag{11}$$

where Δ represents related uncertainty.

The state space representation of system without uncertain parameters can be expressed as:

$$A = \begin{bmatrix} 0 & 1 \\ -1 & -0.2 \end{bmatrix}, \ b = [83 \quad 0], \ C = [1 \quad 0], \ D = 0 \tag{11}$$

By $\gamma = 0.1$ and in the presence of time delay input, the feedback gain is obtained as $K = [-80.94 \ -30.32]$. When the controller is feasible for the maximum time delay τ_1, we can stabilize the system for any delay belonging to the interval $[0 \ \tau_1]$ with H∞ index, γ. The convergence rate without and with time delay is depicted in Figs. 5 and 6, respectively. It should be noticed that the convergence rate during time delay tends toward the maximum delay that by which the system is still stable.

The peak input control ratio, *i.e.* $u_{\max}(\tau)/u_{mx}(0)$ is presented in Fig. 7. It is observed that the ratio increases as the time delay increases, of course with a little fluctuation. The maximum increase in peak occurs when the delay τ is about 120 while the maximum admissible delay τ_1 is obtained 82.5362. In other words, the procedure is valid even if we are in out of the range found for admissible delay. From it, the applicability of the proposed procedure can be concluded.

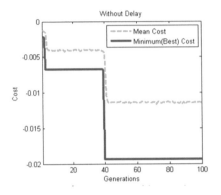

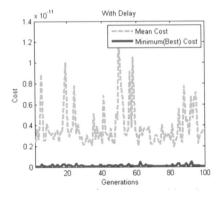

Fig. 5. Rate of convergence without Delay   **Fig. 6.** Rate of convergence with Delay

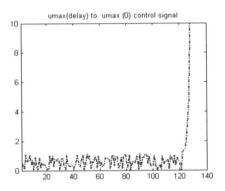

Fig. 7. Ratio of control signal vs. time delay

Table 1 describes the performance of proposed approaches in terms of various factors such as maximum delay, time consuming, gain, etc. The forth row of the table makes a comparison of the four algorithms in term of maximum time delay found by each algorithm, while the first row indicates the worst cost of each algorithm performance. Results show that GA-PSO-ACO finds better solutions by finding more amplitude for maximum delay, yet its worst cost is less than that of others. It should be noticed that GA-PSO-ACO reaches to this result by applying the number of individuals, which obtained by ant colony optimization algorithm. In addition, the required performance time is less than that of others, but not when it is compared to GA-PSO. The performance of GA-ACO in terms of time consuming is poorer than that of GA-PSO but marginally better than that of GA design. It can be concluded that this hybrid method does not lead, actually, to a strictly better performance in comparison to any of the other two. Furthermore, the results of the GA-PSO algorithm show that, it is better than GA and GA-ACO algorithms but still the best performance belongs to GA-PSO-ACO algorithm. It is better at a fraction of the computing time respect to others. Tenth row in Table 1 shows the comparison of different algorithms in overshoot term. The comparison about their settling time is presented in eleventh row of Table 1. These results reveal that the best performance belongs to GA-PSO-ACO and GA-PSO with kp = 50 (*i.e.* fifty percent of population is done by GA algorithm). Although the transient oscillations in GA-PSO and GA-ACO suppressed, they are still inferior to those of GA-PSO-ACO.

Table 1. Comparison of the design results of proposed algorithms

Property	GA (No Delay)	GA (Delay)	GA-ACO (Delay)	GA-PSO (Delay)	GA-PSO-ACO (Delay)
Worst cost (10^{-11})	$-.005 \times 10^{11}$	1.35	1.38	0.89	0.107
Best cost (10^{-13})	$-.0019 \times 10^{13}$	1.25	1.15	1.09	0.679
Gain	(68.2, 52.7)	(−80.94, −30.32)	(−75.38, −80.37)	(1.25, −100)	(28.68, −99.80)
Max delay	–	82.54	81.37	84.51	90.33
Gamma (γ)	0.1	0.1	0.1	0.1	0.1
Time consuming (s)	120	255.39	234.90	92.61	112.90
Best value of $t_{min}(10^{-11})$	$-.162 \times 10^{9}$	2.5	1.88	1.3	1.02
Iteration times	9	15	12	10	8
Overshoot (cm)	.006	.05	.025	.078	.02
Setting time (ms)	12	35	37	28	20

One point that may affect the optimal solution performance and stability is the searching range of output feedback gain. Therefore, for the first execution of algorithm a wider solution space can be used, but after getting the early solution the range can be shorten nearer to the values obtained in the first execution. Table 2 shows the performance of system with considering this point when the initial range is [−100 100]. From which it can be concluded that not only the performance in terms of worst and best cost improves but also it requires lower running time with using this shorten range.

Table 2. Second execution in shorter range

Algorithm	Worst cost (10^{-11})	Best cost (10^{-13})	Time (s)
GA	1.200	6.22	184.98
GA-ACO	1.090	5.10	180.23
GA-PSO	0.324	4.07	78.14
GA-PSO-ACO	0.054	2.41	100.56

6 Conclusion

Lack of sufficient knowledge for modeling system under control often leads to the designing of ill-conditional controller mainly due to simulation flows. This issue is important, especially, when time delay is main concern. Evolutionary algorithms can explore deficiencies greedily. In this paper, an approach has been presented to overcome the stated problem with combining GA and LMI solver for designing a delay dependent H∞ controller. The proposed method employs GA as a searching tool to seek feedback gain to make controller capable of handling non-convex optimization with LMI solver to take stability into account, in the presence of delay. Simulation results confirm the viability of soft proposed methodology by getting admissible input time delay without occurring instability. In addition, in this paper, three combined evolutionary methods on this problem were of interest to accelerate convergence as well as to prevent premature convergence. The results of simulation show the aims are achieved. It was also showed that the algorithm GA-ACO-PSO with using ACO as a tool for tuning GA population size parameter and PSO to enhance part of individuals finds better solution in different terms comparing to GA, GA-ACO, and GA-PSO.

References

1. Zames, G.: Feedback and optimal sensitivity: model reference transformations, multiplicative seminorms, and approximate inverses. IEEE Trans. Autom. Control **26**, 301–320 (1981)
2. Haupt, R.L., Haupt, S.E.: Practical genetic algorithms (2004)
3. El-Razaz, Z.S., Mandor, M.D., Ali, E.S.: Damping controller design for power systems using LMI and GA Techniques. In: Eleventh International Middle East Power Systems Conference, MEPCON 2006, pp. 297–301 (2006)
4. Chung, H.-Y., Wu, S.-M., Yu, F.-M., Chang, W.-J.: H2/ H∞ Robust static output feedback control Design via Mixed genetic algorithms and linear matrix inequility. J. Dyn. Syst. Meas. Control. ASME **127**, 715–722 (2005)
5. Su, T.-J., Wei, C.-P., Jong, G.-J.: FIR equalization for nonlinear communication channels. In: American Control Conference, p. 6 (2006)
6. de Araujo, H.X., Garbin Langner, C.: H∞ control for uncertain systems under time domain constraints. In: 44th IEEE Conference on Decision and Control, 2005 and 2005 European Control Conference, CDC-ECC 2005, pp. 1325–1330 (2005)
7. Du, H., Shi, X.: Low-Order H∞ controller design using LMI and genetic algorithm. In: Proceedings of the 2002 American Control Conference, vol. 1, pp. 511–512 (2002)
8. Hung, J.-C., Chen, B.-S.: Genetic algorithm approach to fixed-order mixed H2/ H∞ optimal deconvolution filter designs. IEEE Trans. Signal Process. **48**, 3451–3461 (2000)
9. Dyn, J.: LMI based to robust genetic control design for a class of uncertain neutral systems with state and input delay. J. Dyn. Syst. Meas. Contr. **128**, 675–681 (2006)
10. Du, H., Lam, J., Sze, K.Y.: H∞ disturbance attenuation for uncertain mechanical systems with input delay. Trans. Inst. Meas. Control. **27**, 37–55 (2005)
11. Du, H., Zhang, N.: Control for buildings with time delay in control via linear matrix inequalities and genetic algorithms. Eng. Struct. **30**, 81–92 (2008)
12. Lee, Z.-J., Su, S.-F., Chuang, C.-C., Liu, K.-H.: Genetic algorithm with ant colony optimization (GA-ACO) for multiple sequence alignment. Appl. Soft Comput. **8**, 55–78 (2008)

13. Kaveh, A., Shahrouzi, M.: A hybrid ant strategy and genetic algorithm to tune the population size for efficient structural optimization. J. Eng. Comput. **24**, 237–254 (2007)
14. Silva, C.A., Faria, J.M., Abrantes, P., Sousa, J.M.C., Surico, M., Naso, D.: Concrete Delivery using a combination of GA and ACO. In: 44th IEEE Conference on Decision and Control, 2005 and 2005 European Control Conference, CDC-ECC 2005, pp. 7633–7638 (2005)
15. Dorigo, M., Stutzle, T.: Ant Colony Optimization (2004)
16. Lu, C.-F., Juang, C.-F.: Evolutionary fuzzy control of flexible AC transmission system. IEE Proc. Gener. Transm. Distrib. **152**, 441–448 (2005)
17. Kennedy, J., Eberhart, R.C.: Swarm Intelligence. Morgan Kaufmann, San Francisco (2001)
18. Gahinet, P.: Explicit controller formulas for LMI-based H∞ synthesis. In: American Control Conference, vol. 3, pp. 2396–2400 (1994)
19. Xiong, W.-Q., et al.: An improved real-code genetic algorithm. In: Proceedings of 2004 International Conference on Machine Learning and Cybernetics, vol. 4, pp. 2361–2364 (2004)
20. Gahinet, C.P., Nemirovski, A., Laub, A.J., Chilali, M.: LMI Control Toolbox. ed: The MathWorks (2008)
21. Alfassio Grimaldi, E., Gandelli, A., Grimaccia, F., Mussetta, M., Zich, R.E.: A new hybrid technique for the optimization of large –domain electromagnetic problem. In: Proceedings of URSI GA (2005)

Sharing Without Losing and Donation: Two New Operators for Evolutionary Algorithm with Variable Length Chromosome

Rajesh Reghunadhan[1,2P(✉)]

[1] Department of Computer Science, Central University of South Bihar,
Patna 800014, Bihar, India
kollamrajeshr@ieee.org
[2] Department of Computer Science, Central University of Kerala, Tejaswini Hill,
Periya, Kasaragod 671316, Kerala, India

Abstract. Evolutionary algorithm (EA) is considered as a simple, powerful and derivative free optimization technique derived from the concepts of nature with capabilities of parallel implementation. Many researchers have made use of variable length chromosome (VLC) and its related operators in Evolutionary optimization. It can be seen that many organisms survive by donating their food. If one donates some of his food, then he loses some of his food. A new operator is proposed based on donation of food. In the proposed donation operator, if a chromosome donates some of its genes to other chromosome then it loses some of the genes and the other chromosome which gets these genes will have an increase in the number of genes. It can also be seen from the nature that many organisms survive by sharing their knowledge and while sharing there is no loss on both sides. Based on the knowledge sharing, a new operator is proposed which shares the genes between two selected chromosomes thereby not losing any individual genes but gaining additional genes with increase in the length of the chromosomes. Evolutionary algorithm with the proposed operators is used to optimize fuzzy rules and the results prove that the new operators have merits.

Keywords: Evolutionary algorithm · Variable length chromosome
Sharing without losing · Donation

1 Introduction

Natural evolution and Darwin's theory of natural optimization are the motivation for the discovery of new operators and with novel functionalities in a variety of algorithms including evolutionary algorithms and its flavours like genetic algorithm [1] and evolutionary strategies [2]. Hence most of the rules and theories associated with natural evolution also apply to these computational algorithms. Moreover, these computational algorithms, including evolutionary algorithms are simple, powerful, general purpose, derivative free optimization techniques which are subjected to crossover and mutation in a selective environment where only the fittest will survive with capabilities of considering the points simultaneously in parallel [3].

© Springer Nature Switzerland AG 2019
B. K. Ane et al. (Eds.): WSC 2014, AISC 864, pp. 158–166, 2019.
https://doi.org/10.1007/978-3-030-00612-9_14

Evolutionary algorithms work on the principle of five different operators namely, fitness evaluation, selection, crossover, mutation, and reinsertion [9].

The fitness evaluation of each chromosome in the population is performed using a fitness function. The fitness function is different for different problem and there can be more than one fitness function for a problem. Choosing the best fitness function for the given problem is again a research issue.

There are various selection mechanisms [10] available, namely, roulette wheel selection, rank based selection, tournament based selection, etc. The main idea behind the selection mechanism is that, generation after generation the average fitness should increase after selection of the chromosomes.

There are various crossover operators, namely, one-point crossover, two-point crossover, multi-point crossover, shuffle crossover, etc. In these crossovers, the genes of the chromosomes are exchanged at the position of the crossover points. There are also crossovers which creates offspring based on the linear combination of the two parent genes, for example the arithmetic crossover [11]. Other crossover includes, ripple crossover for grammatical evolution [12], Eigen vector based crossover operator [13], chemical crossover [14], etc.

Mutation operators randomly make changes in the values of the genes in random positions with a probability of mutation and based on a mutation threshold.

Most of the evolutionary algorithms in literature start with random initialization of population. The chromosomes in the population are of fixed size and the length of the chromosome never changes with respect to operators. Recently, evolutionary algorithm with variable length chromosome are being studied in the literature [4–8, 15]. These variable length chromosomes require extended versions of the crossover and mutation operators.

It can be seen that many organisms survive by donating their food. If one donates some of his food, then he loses some of his food. A new operator is proposed based on donation of food. In the proposed donation operator, if a chromosome donates some of its genes to other chromosome then it loses some of the genes and the other chromosome which gets these genes will have an increase in the number of genes.

It can also be seen from the nature that many organisms survive by sharing their knowledge and while sharing there is no loss on both sides. Based on the knowledge sharing, a new operator is proposed which shares the genes between two selected chromosomes thereby not losing any individual genes but gaining additional genes with increase in the length of the chromosomes. This operator is very different from other crossovers and it can be seen that there is no loss in the number of genes/features.

Evolutionary algorithm with the proposed operators is used to optimize fuzzy rules and the results prove that the new operators have merits. This paper is organised as follows. Section 2 deals with the new proposed operators. Section 3 deals with the results and discussion and Sect. 4 concludes the paper.

2 Sharing Without Losing and Donation for EA with VLC

2.1 Donation

The real survival of natural organisms depends on sharing and donation of food, water and air. If an organism, say X gives/donates some of its food to another organism, say Y, then X loses some of its food and at the same time Y gains some food. The same is the case if both X and Y try to donate some of their food where both X and Y gains some new food by losing some of its food. The first new operator is proposed based on this donation scheme. Some of the applications of these donation schemes includes (i) travel sales man problem where the length of the route in each chromosome is different, (ii) circuit design, (iii) fuzzy rule optimization, (iv) optimizing the number of clusters, etc.

One side donation: In one side donation, one of the parent chromosomes donates its gene to other parent chromosome forming offspring chromosomes. For example, Fig. 1 shows the parent chromosomes and Fig. 2 shows the offspring chromosomes. It can be seen that the second parent chromosome donates some of its fruits to first chromosome forming two offspring, where the first offspring has some gains with increase in length while the second chromosome has some decrease in length due to gene losing.

Fig. 1. Parent chromosomes

Fig. 2. Offspring chromosomes after donation

Fig. 3. Parent chromosomes

Two-side donations: Two-side donation is equivalent to extended crossover with different crossover points (since the length of the parent chromosome are different) for each parent chromosomes. For example, Figs. 4 and 5 shows the two cases (possibilities) of two-side donations between two parent chromosomes (Fig. 3) with donation points dp1 and dp2. This two-side donation can be extended for multiple donation points, shuffle donation, etc.

Fig. 4. Case1: Offspring chromosomes after donation

Fig. 5. Case2: Offspring chromosomes after donation

2.2 Sharing Without Losing

The natural optimization and survival also depends on knowledge sharing between different organisms. The 2nd new operator that is proposed in this paper is based on knowledge sharing and hence by sharing knowledge no organism loses its knowledge but gains additional knowledge. Some of the applications of these sharing schemes includes (i) travel sales man problem where the length of the route in each chromosome is different, (ii) circuit design, (iii) fuzzy rule optimization, (iv) optimizing the number of clusters, etc.

One side sharing: In one side sharing one of the parent chromosomes try to share its genes to the other chromosome without losing any genes. For example, Table 1 shows the one side sharing, where the parent1 shares some of it rules to parent2 forming two offspring, namely, offspring1 and offspring2. It can be seen that there is no loss in the number of genes.

Table 1. One Side Sharing without losing

Parent1	if x > 2 then y = 4	if x < −2 then y = 0	
Parent2	if x > 0 then y = 4 + x	if x < 0 then y = 4−x	
offspring1	if x > 2 then y = 4	if x < −2 then y = 0	
offspring2	if x > 0 then y = 4 + x	if x < 0 then y = 4−x	if x > 2 then y = 4

Two side sharing: In two side sharing, the parent chromosomes share some of their genes without losing any genes. For example, Table 2 shows the two-side sharing, where the parent1 and parent2 shares some of their rules forming two offspring, namely, offspring1 and offspring2. It can be seen that there is no loss in the number of genes. This two-side sharing can be extended for multiple sharing points, shuffle sharing, etc.

Table 2. Two Side Sharing without losing

Parent1	if x > 2 then y = 4	if x < –2 then y = 0	
Parent2	if x > 0 then y = 4 + x	if x < 0 then y = 4–x	
offspring1	if x > 2 then y = 4	if x < –2 then y = 0	if x < 0 then y = 4–x
offspring2	if x > 0 then y = 4 + x	if x < 0 then y = 4–x	if x > 2 then y = 4

2.3   Other Operators

Gene deletion (Pruning): Genes will be deleted randomly from a chromosome with a probability GDP. There will also be a threshold (indicative parameter about the number of genes) above which the genes should be deleted.

Gene Inclusion (Feeding): Genes will be included in a chromosome with a probability GIP. There will also be a threshold (indicative parameter about the number of genes) below which the genes should be included.

Additional Operators: If the number of genes exceeds the maximum limit, then there is an operator which randomly deletes genes. Moreover due to crossover and mutation, if some genes are repeated then there is an operator which deletes those genes which are repeated.

3   Results and Discussion

3.1   Design and Stability Analysis of TSFC Using EA with VLC

The T-S fuzzy controllers are designed based on the assumption of the existence of T-S fuzzy model of the system with antecedent part of the rules fixed by experts and then design the gain values or consequent parts by using linear matrix inequalities or by evolutionary methods.

In this scenario for the design of antecedent part, variable length chromosomes can be used and consequent part can be designed using linear matrix inequality. Here the chromosomes will contain m genes, where m is equal to the number of input variables. Each gene, g_i corresponding to the variable x_i, and can have L_i alele. Each of the alele will represent the centres of the membership functions for the variable by assuming constant or appropriate spread to cover the domain of the variable. Since the genes are of different length, the new operators can be used for the evolution of the rules. The fitness function can also take care about the stability constraint for the design of stable controller. More detail about the design and stability analysis of Takagi-Sugeno fuzzy controller with variable length chromosome and its results can be had from our paper (Rajesh et al. [16]).

3.2   Design of Mamdani Type Fuzzy Controllers Using EA with VLC

In order to efficiently sample the search space, to optimize the number of rules, and to optimize the parameters of both antecedent and consequent fuzzy terms of Mamdani type rules, variable length chromosome can be used. Here i^{th} chromosome will contain

m_i genes, where each gene corresponds to a rule. Each gene/rule will intern have antecedent part and consequent part. Both the antecedent part and consequent part will intern contain centre and spread of the fuzzy terms or the membership functions. More detail about the design of Mamdani type fuzzy controller using variable length chromosome can be had from our paper (Rajesh et al. [17]).

3.3   Rule Selection Using EA with VLC

In this paper, rule selection for fuzzy logic controller is done using evolutionary algorithm with variable length chromosome to show the effectiveness of the proposed operators.

Benchmarking control problem, namely, the inverted pendulum is used for the study. Seven membership functions are designed for the input variable, angle (θ) in the range $[-75°, 75°]$ and is shown in Fig. 6. Thirty membership functions are designed for the angular velocity $(d\theta/dt)$ in the range $[-1000, 1000]$ and is shown in Fig. 7. The output variable, force takes 21 singleton values, namely, $-100, -90, -80, -70, -60, -50, -40, -30, -20, -10, 0, 10, 20, 30, 40, 50, 60, 70, 80, 90, 100$. Based on the membership functions, $7 \times 30 \times 21 = 4410$ rules are designed. Now it is the duty of the evolutionary algorithm to choose the minimum number of best rules from 4410 rules for the fuzzy logic controller.

The chromosome is designed to have n number of genes. Each gene will correspond to a rule number. Initial population with random rules are created. Five different simulations are carried out starting with initial number of rules as 1, 100, 25, 50 and 75.

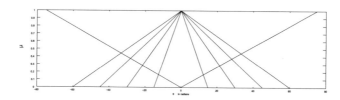

Fig. 6. Seven Membership functions for theta. [leftspread,centre,rightspread] are [Inf, −75, 0], [15, 0, 15], [75, 75, Inf], [30, 0, 30], [45, 0, 45], [60, 0, 60]

The parameters of evolutionary algorithm are provided as, (1) population size = 100, (2) gene mutation rate = 0.2, (3) new gene inclusion rate = 0.2, (4) gene deletion rate = 0.2, (5) new gene inclusion threshold = 20 rules, (6) gene deletion threshold = 5 rules, (7) Donation rate = 0.20, (8) Two side sharing rate = 0.20, (9) One side sharing rate = 0.2, (10) Gene exchange/crossover rate = 0.2, (11) Maximum Generation allowed = 100, (12) Maximum number of rules allowed = 100.

Fitness of the best chromosome in each generation and the number of rules of the best chromosome in each generation are shown in Figs. 9 and 10 respectively for 5 different cases of simulation. Figure 8 shows the pendulum angle for the simulation with the chromosome obtained after 50th generation starting with 100 rules. The results show the merits of the rule optimization capability of the operators.

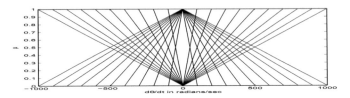

Fig. 7. Thirty Membership functions for $(d\theta/dt)$. [leftspread,centre,rightspread] are [100, 0, 100], [200, 0, 200], [300, 0, 300], [400, 0, 400], [500, 0, 500], [600, 0, 600], [700, 0, 700], [800, 0, 800], [900, 0, 900], [1000, 0, 1000], [∞, −100, 100], [∞, −200, 200], [∞, −300, 300], [∞, 400, 400], [∞, −500, 500], [∞, −600, 600], [∞, −700, 700], [∞, −800, 800], [∞, −900, 900], [∞, −1000, 1000], [100, 100, ∞], [200, 200, ∞], [300, 300, ∞], [400, 400, ∞], [500, 500, ∞], [600, 600, ∞], [700, 700, ∞], [800, 800, ∞], [900, 900, ∞], [1000, 1000, ∞]

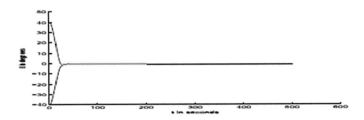

Fig. 8. Performance of the best chromosome after 50 generations starting with 100 rules

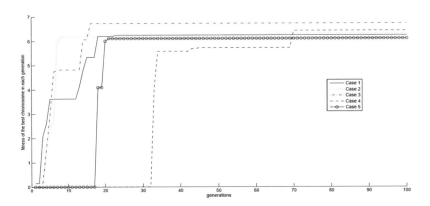

Fig. 9. Fitness of the best chromosome in each generation for simulations starting with 1 rule (Case 1, straight line), 100 rules (Case 2, dotted line), 25 rules (Case 3, dash dot), 50 rules (Case 4, dash), 75 rules (Case 5, circle)

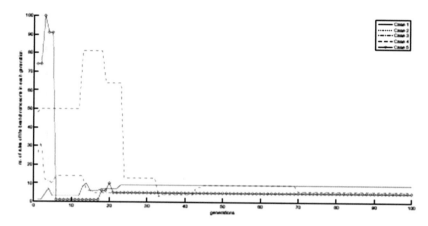

Fig. 10. No of rules of the best chromosome in each generation for simulations starting with 1 rule (Case 1, straight line), 100 rules (Case 2, dotted line), 25 rules (Case 3, dash dot), 50 rules (Case 4, dash), 75 rules (Case 5, circle)

4 Conclusion

Evolutionary algorithm with the proposed operators is used to optimize fuzzy controller rules and the results prove that the new operators have merits. Future research is going on for the design and optimization of travelling salesman problem.

References

1. Harik, G.R., Lobo, F.G., Goldberg, D.E.: The compact genetic algorithm. IEEE Trans. Evol. Comput. **3**(4), 287–297 (1999)
2. Francois, O.: An evolutionary strategy for global minimization and its markov chain analysis. IEEE Trans. Evol. Comput. **2**(3), 77–90 (1998)
3. Alba, E., Tomassini, M.: Parallelism and evolutionary algorithms. IEEE Trans. Evol. Comput. **6**(5), 443–462 (2002)
4. Kim, I.Y., Weck, O.L.: Variable chromosome length GA for progressive refinement in topology optimization. Struct. Multidisc. Optim. **29**, 445–456 (2005)
5. Kim, I.Y., Weck, O.L.: Variable chromosome length genetic algorithm for structural topology design optimization. In: Proceedings of 45th AIAA-ASME-ASCE-AHS-ASC Structures, Structural Dynamics & Materials Conference, California, April (2014)
6. Khan, R.Z., Ibraheem, N.A.: Genetic shape fitting for hand gesture modeling and feature extraction using VLC. Br. J. Sci. **10**(1) (2013)
7. Cavill, R., Smith, S.L., Tyrrell, A.M.: Variable length GA with multiple chromosomes on a variant of the Onemax problem. GECCO, USA, pp. 1405–1406 (2006)
8. Zhang, M., Deng, Y., Chang, D.: A novel genetic clustering algorithm with VLC representation. GECCO, Canada, pp. 1483–1484 (2014)
9. Schmitt, L.M.: Theory of genetic algorithms. Theoret. Comput. Sci. **259**, 1–61 (2001)
10. Rogers, A., Prugel-Bennett, A.: Genetic drift in genetic algorithm selection schemes. IEEE Trans. Evol. Comput. **3**(4), 298–303 (1999)

11. Yalcinoz, T., Altun, H., Uzam, M.: Economic dispatch solution using a genetic algorithm based on arithmetic crossover. In: IEEE Porto Power Tech Proceedings, vol. 2 (2001)
12. Oneill, M., Ryan, C., Keijzer, M., Cattolico, M.: Crossover in grammatical evolution. Genet. Program Evolvable Mach. **4**, 67–93 (2003)
13. Guo, S.-M., Yang, C.-C.: Enhancing differential evolution utilizing eigenvector based crossover operator. IEEE Trans. Evol. Comput. **19**(1), 31–49 (2014)
14. Bersini, H.: The immune and the chemical crossover. IEEE Trans. Evol. Comput. **6**(3), 306–313 (2002)
15. Hutt, B., Warwick, K.: Synapsing variable-length crossover: meaningful crossover for variable-length genomes. IEEE Trans. Evol. Comput. **11**(1), 118–131 (2007)
16. Rajesh, R., Kaimal, M.R.: GAVLC- GA with VLC for the simultaneous design and stability analysis of T-S fuzzy controllers. In: IEEE International Conference on Fuzzy Systems, WCCI 2008, pp. 1389–1396 (2008)
17. Rajesh, R., Kaimal, M.R.: GAVLCRG- Genetic algorithm with variable length chromosome-based rule generation scheme for fuzzy controllers. Adv. Fuzzy Sets Syst. **4**(1), 33–66 (2009)
18. Cordon, O., Gomlde, F., Herrera, F., Hoffmann, F., Magdalena, L.: Ten years of genetic fuzzy systems: current framework and new trends. Fuzzy Sets Syst. **141**, 5–31 (2004)
19. Ishibushi, H., Nozaki, K., Yamamoto, N., Tanaka, H.: Selecting fuzzy if-then rules for classification problems using genetic algorithm. IEEE Trans. Fuzzy Syst. **3**(3), 260–270 (1995)
20. Lee, M., Takagi, H.: Integrating design stages of fuzzy systems using genetic algorithms. In: Proceedings of 2nd IEEE International Conference on Fuzzy Systems, San Francisco, CA, pp. 612–617 (1993)
21. Setnes, M., Roubos, H.: GA-fuzzy modeling and classification: complexity and performance. IEEE Trans. Fuzzy Syst. **8**(5), 509–522 (2000)

Optimizing Quantitative and Qualitative Objectives by User-System Cooperative Evolutionary Computation for Image Processing Filter Design

Satoshi Ono[✉], Hiroshi Maeda, Kiyomasa Sakimoto, and Shigeru Nakayama

Department of Information Science and Biomedical Engineering,
Graduate School of Science and Engineering, Kagoshima University,
1-21-40, Korimoto, Kagoshima 890-0065, Japan
{ono,sc105050,sc108025,shignaka}@ibe.kagoshima-u.ac.jp

Abstract. This paper proposes a cooperative optimization method between a system and a user for problems involving quantitative and qualitative optimization criteria. In general Interactive Evolutionary Computation (IEC) models, a system and a user have their own role of evolution, such as individual reproduction and evaluation. In contrast, the proposed method allows them to dynamically switch their roles during the search by using explicit fitness function and case-based user preference prediction. For instance, in the proposed method, the system performs a global search at the beginning, the user then intensifies the search area, and finally the system conducts a local search at the intensified search area. This paper applies the proposed method for an image processing filter design problem that involves both quantitative (filter output accuracy) and qualitative criterion (filter behavior). Experiments have shown that the proposed cooperation method could design filters that are in accordance with user preference and have better performance than filters obtained by Non-IEC search.

1 Introduction

Designing a fitness function is crucial to find good solutions by Evolutionary Computation (EC) algorithms in real world problems. Ideal function may not be sufficient to solve such real world problems, and ambiguous complemented subfunctions may be necessary to obtain practical quasi-optimal solutions. Such additional criteria are not always apparent in advance, and it is intractable to redesign the fitness function during the search.

Interactive Evolutionary Computation (IEC) is a way to solve problems in which implicit human preference and emotion are necessary to evaluate solutions (Takagi 2001). In IEC, solution evaluation is performed by a human user, and explicit fitness functions are not used in general IEC models. Various IEC applications have been proposed such as computer graphics (Huang 2006; Kim 2000; Unemi 1998), musical composition (Ando 2007; Unemi 2001), and so on.

© Springer Nature Switzerland AG 2019
B. K. Ane et al. (Eds.): WSC 2014, AISC 864, pp. 167–178, 2019.
https://doi.org/10.1007/978-3-030-00612-9_15

Real world problems require being optimized based on both quantitative and qualitative criteria as mentioned above. Although some studies have conducted to estimate user preference to reduce user fatigue in evaluating solutions in IEC (Amamiya 2009; Osaki 1998), research for optimization based on both qualitative and quantitative optimization criteria has just increased (Hiroyasu 2010).

This paper proposes a cooperative EC method between a system and its user for problems involving qualitative and quantitative optimization criteria. The method allows a user to dynamically switch the role of EC at any time during the search. To achieve this, the proposed method estimates user preference by using Case-Based Reasoning (CBR) (Riesbeck 1989; Schank 1994). For instance, the proposed method allows a user to evaluate solutions or to do a genetic operation at any time the user wants to do them. The model also allows a user to carefully redesign solutions or to evaluate them at every generation.

To verify the effectiveness of the proposed user-system cooperative EC, this paper focuses on image filter generation that is a problem to approximate an unknown filter by a combination of known primitive filters. This problem involves both qualitative and quantitative criteria, which correspond to filter quality and behavior respectively. Users who are familiar with image processing can edit the solution structure directly, and other users can choose appropriate solutions by watching the output images. Experiments have shown that the proposed method could respond various user demands for cooperative search, and that the cooperation brought image filters that have competitive performance and different structure from filters obtained by Non-IEC search.

2   Image Filter Generation

Image processing is one of the most rapidly evolving areas of information technology today, with growing applications in various areas of science and engineering. Although many advanced technologies and research studies in image processing have reached to being practical, trial and error are still necessary to design an image processing program even for professionals.

Various methods have been proposed for improvement and automation of image processing algorithm design. Nagao has proposed EC-based methods that approximate an unknown image filter by a combination of known primitive filters connected in a list, a tree or a graph (Nagao 1996; Shirakawa 2007). The filter structure is obtained by evolutionary computation algorithms such as Genetic Algorithm (GA), Genetic Programming (GP) and so on. The methods do not require human experts' help and interaction, and is applicable to various image processing filter generation problems.

Nagao's methods produce plural filters having almost the same fitness value but different image processing characteristics in problems in which it is hard for the methods to generate an optimal filter. For instance, in a traffic sign extraction problem (Maeda 2010; Maezono 2006), it is quite hard to generate an ideal filter which extracts traffic signs even in back-lighting or occlusion environments and prevents miss detection of background areas having colors similar to the signs.

Therefore, the methods generate various semi-optimal filters which have similar fitness value and different image processing behavior; some filters avoids noise but sometimes fails to find signs, and other filters do not miss the signs but outputs noisy results. In practical use, a user must choose the most appropriate one from the semi-optimal filters, but the method may not produce a filter adequate for the user's purpose. Although users demand should be incorporated into a fitness function of the method, it is hard to predict the output filters' behavior by considering the method and the problems property in advance.

In addition, there is a demand from professionals in image processing and computer vision for filter design methods to introduce their preference and heuristics into the search. The professionals want to edit solutions when they think of a good filter structure.

3 The Proposed Method

3.1 The Basic Ideas

The aim of this paper is to produce a user-system collaborative EC method for simultaneous optimization of quantitative and qualitative objective functions (Table 1). The principles of the proposed method are as follows:

1. **IEC and Non-IEC search are switchable at any time:** The proposed method basically performs non-IEC search and displays the obtained solutions for every generation. The user can stop the search whenever the user wants to evaluate or edit solutions. Non-IEC search of the proposed method can be performed by any multi-point search algorithm. In this paper, Genetic Algorithm (GA) with Minimal Generation Gap (MGG) model is adopted because its behavior is analyzed and clarified (Satoh 1996).
2. **Explicit fitness function:** In an image processing filter generation problem, a solution must be evaluated from output image quality. Therefore explicit fitness function is defined and used, whereas general IEC models do not. User evaluation score is involved in the fitness function.
3. **User preference estimation by CBR:** For user preference estimation with the least user fatigue of solution evaluation, the proposed method uses Case-Based Reasoning (CBR) model (Riesbeck 1989; Schank 1994). A solution candidate and its evaluation by the user are stored as a case into a case base, and the case is used to estimate user preference value of other unevaluated candidates.
4. **Network-structured image filter:** The proposed method generates network-structured image filters based on Genetic Image Network (GIN) (Shirakawa 2007), which is an automatic construction method of graph-structured image processing algorithm. GIN generates filters involving loop and feedback structures which enable to partially reuse image processing results. Hence, the proposed method generates more compact filters than tree-structured ones.

Table 1. The relationship between the proposed method and general IEC and Non-IEC.

	Non-IEC	IEC	Proposed user-system cooperative method
Objective function	Quantitative	Qualitative	Quantitative and Qualitative
User operation	None	Evaluation at each generation	Evaluation and direct edit at any time

The proposed method based on the above ideas enables a cooperative search by a system and a user. Example user-system cooperation ways by the proposed method are as follows (Ono and Nakayama 2010):

(a) **Manual mode, later Non-IEC mode**
At the beginning of the search, the user designs an outline of filter structure based on his/her knowledge and heuristics, and then the system searches for detailed filter structure and thresholds of primitive filters. The former user operation is regarded as a global search which guides to the search space in which appropriate solutions exist, and the latter system search as a local search.

(b) **IEC mode, later Non-IEC mode**
The early stage of the search, the user looks for adequate solutions for the user's purpose by evaluating solutions generated by the system. After the system gets to produce solutions in accordance with user demand by case accumulation, the system searches better solutions similar to (a).

(c) **Non-IEC mode, later IEC and/or Manual mode**
The early stage of the search, the system searches solutions by itself and the user observes the solutions to derive inspiration from them as shown in Fig. 1. After the system produces some promising solution candidates, the user chooses adequate ones from them or revises one of them in order to enhance search intensification. Non-IEC mode search may improve the solution chosen or revised by the user. This way is regarded as a combination of a global search by the system and a local search by the user.

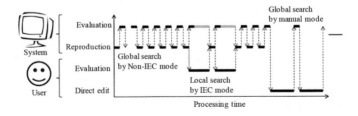

Fig. 1. Example of user-system collaboration in the proposed method.

3.2   Codification

A solution in the proposed method is a network-structured image processing filter. To solve this problem with GA, genotype and phenotype of a solution are represented as Fig. 2 based on the previous work (Shirakawa 2007). Chromosome consists of a sequence of pairs involving two types of genes: primitive filter identification number and a list of input edges, as shown in Fig. 2(a). A pair of genes in genotype is necessary to compose a node and edges connected to the node in a graph. Phenotype is a graph which consists of nodes corresponding to a primitive filter and edges between nodes. Each primitive filter affects an image given from a previous node connected by an incoming edge, and sends its output image to next nodes connected by an outgoing edge. There is a timescale on a filter graph; all nodes works once a step, the filter graph stops at the number of steps determined beforehand, and outputs an image of an output filter node at that step.

Primitive filters are categorized into three groups according to the default input edge number: primitive filters without input edge, ones with an input edge, and ones with two input edges. Primitive filters without input edge outputs color components of an input image such as RGB and HSB components. Primitive filters with an input edge are simple, well-used image processing filters such as edge extraction, embossing, binarization, inverse and so on. Primitive filters with two input edges figure out an output image by the specified operator such as subtraction, addition, average and so on. A node can have more input edges than the number of default edges of its primitive filter; for instance, a node having smoothing filter can have more than one input edge, and a node of average filter more than two input edges. In such cases, input edges are prioritized in the reverse chronological order of input image update. In opposite, a primitive filter whose default number of input edges is two outputs the same image as an input image when the primitive filter has just one input edge.

Some of primitive filters have their internal parameters; for instance, binarization thresholds, multiplication value, and addition value. Such internal parameters start their initial values and modified by mutation.

A gene corresponding to a primitive filter node consists of a set of an integer value representing a primitive filter identification number and a real value representing an internal filter parameter, and a gene corresponding to an edge list is represented as a variable length list of integer values.

3.3   Algorithm

Figure 2(c) shows the structure and the process flow of the proposed method. The proposed method conducts Non-IEC search and suggestion of solutions (individuals) unless a user stops the search. The proposed method uses a case base to evaluate individuals in Non-IEC search if cases are stored in the case base.

A user observes presented individuals and stops the search whenever the user finds a favorite individual or whenever no individuals suit the users demand. After the search stopped, the user gives an evaluation value to one, some or all

Satoshi Ono, Hiroshi Maeda, Kiyomasa Sakimoto and Shigeru Nakayama

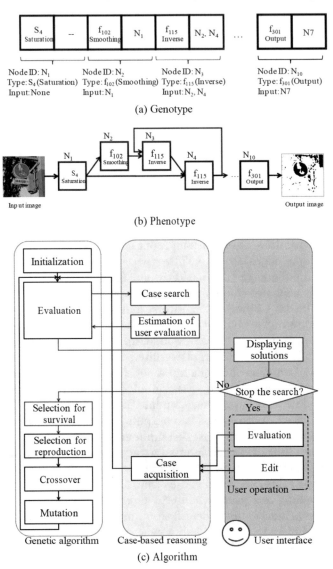

Fig. 2. Codification and algorithm of the proposed method.

individuals, the user resumes the search. Individuals that are evaluated or edited by the user are stored as cases into the case base, and reused later in the search.

In this paper, MGG model is adopted as mentioned above. In selection for reproduction, two parents are selected randomly, and, in selection for survival, two individuals are selected from two parents and two offsprings: the best individual and an individual selected by a roulette selection. Uniform crossover and simple point mutation are adopted in the proposed method.

Figure 3 shows an example screenshot of the system that is an implementation of the proposed method. The system interface consists of two windows: a viewer and an editor. On the viewer, The user can evaluate an individual and mark it within five-level rates by moving a slider shown in the individual display area. On the editor, the user can modify filter graph structure: add, delete or change

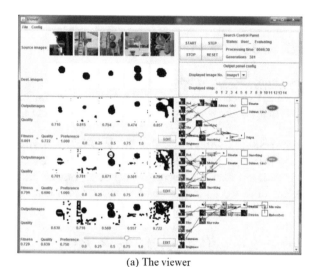

(a) The viewer

(b) The editor

Fig. 3. User interface of the proposed method.

an edge between nodes, and change primitive filter type and inner parameter value.

3.4   Fitness Calculation with User Evaluation Estimation

The proposed method calculates fitness $F(C)$ of an individual C from output image quality $Q(C)$, user evaluation value $P(C)$, and sharing score $S(C)$.

$$F(C) = \left\{ (1 - w^{(p)}) \cdot Q(C) + w^{(p)} \cdot P(C) \right\} \cdot S(C) \qquad (1)$$

$F(C)$, $Q(C)$, $P(C)$, and $S(C)$ are real values ranged from 0 to 1. $w^{(p)}$ is a weight parameter.

In image processing design problem, sets of learning images are given. Each set consists of source, target and weight configuration images. Image processing filter quality $Q(C)$ is an average correspondence rate between output and target images.

User evaluation rank is divided into five levels. $P(C)$ of individuals C which is evaluated by the user is calculated by normalizing the ranked level into $[0, 1]$ value. By CBR, estimated user preference $\overline{P}(C)$ of an individual C is calculated from the preference value $P(C')$ of the nearest case C' and the distance $d(C, C')$ between C and C' by the following equation:

$$\overline{P}(C) = \max\{(T_p - d(C, C')), 0\} \times \frac{1}{T_p} \times (P(C') - 0.5) + 0.5 \qquad (2)$$

Distance between individual C and case C' is defined by the average correspondence rate of pixel values of their output images.

To avoid premature convergence, the proposed method utilizes a niching method by fitness sharing (Horn 1997). Sharing score $S(C)$ of an individual C is calculated by the following equation:

$$S(C) = \begin{cases} \dfrac{1}{|S_{C_{sim}}| - T_{S_{sim}} + 1} & \begin{array}{l} \text{if } C \text{ is not the best} \\ \text{and } |S_{C_{sim}}| > T_{S_{sim}} \end{array} \\ 1 & \text{otherwise} \end{cases} \qquad (3)$$

where $S_{C_{sim}}$ is the set of similar individuals C_j within a radius of T_{sh}, and $|S_{C_{sim}}|$ is the number of the similar individuals. $T_{S_{sim}}$ and T_{sh} are thresholds.

4   Experiments

To verify the effectiveness of the proposed user-system cooperative method, we conduct an experiment with a fruit detection problem in which target image areas are some fruits and vegetables such as a tomato and a banana, as shown in Fig. 4(a) and (b). In this problem, it is difficult to design a filter which can extract all fruit image areas simultaneously. This is because fruit items have different

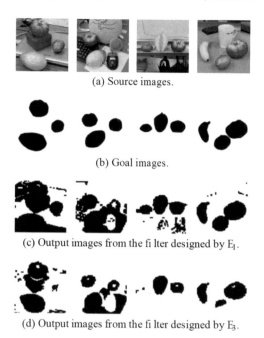

(a) Source images.

(b) Goal images.

(c) Output images from the filter designed by E_1.

(d) Output images from the filter designed by E_3.

Fig. 4. A tested fruit detection problem involving four images

Table 2. Detected rates for each fruit items.

	Image 1			Image 2			Image 3			Image 4			Total	
	Tom	Lim	Pap	Tom	Pea	Pap	Lim	Sta	Tom	Ban	Lim	Tom	All	Preferred
Proposed	20%	70%	30%	80%	20%	10%	80%	90%	90%	100%	90%	80%	63%	85%
Non-IEC	0%	40%	0%	20%	0%	10%	30%	90%	80%	100%	60%	80%	43%	–

colors, and some other objects such as a red box, a yellow mouse pad and a brown desk which have similar colors to the targets are under the target fruit items. Examinees chose a fruit in advance, and designed a filter which mainly finds the chosen fruit in preference to other fruits. Therefore, in this experiment, accuracy of extracted image area is a quantitative objective, and finding the chosen fruit item is a qualitative objective. A target object was regarded as successfully detected if more than 1/3 size of the area was extracted, more than 1/2 of its boundary was clarified, and the extracted area is in one of the top five largest areas extracted.

The proposed method was compared to Non-IEC search which was based on previous work (Shirakawa 2007). Parameteres are configured as follows; population size was set to 50, crossover and mutation rates to 0.5 and 0.03. A number of nodes were fixed to 15. Output image of the output node at 15 step was regarded as an output of the filter. $w^{(p)}$, T_{sim}, and T_{sh} were set to 0.3, 2, and 0.1. The experiments were conducted with PC/AT compatible computers (Intel Core 2 Quad 2.66 GHz, 4 GB RAM). The search time limit was set to 20 minutes.

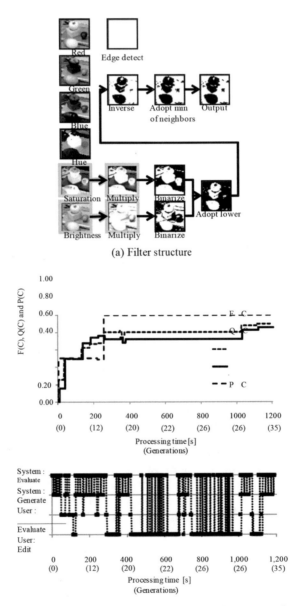

(a) Filter structure

(b) Transitions of fitness and search role between the system and user.

Fig. 5. Examples of filter structure and transitions of fitness and search role between the system and examinees E3 in a fruit detection problem.

Table 2 shows the success rates in detection of the fruit items (tom.: tomato, lim.: lime, pap.: papaya, pea.: peach, sta.: star fruit) averaged over ten examinees and ten runs of Non-IEC search. The proposed user-system cooperative method could find 63% fruit items in average, whereas Non-IEC search could find 43% of them. This shows that users' heuristics helped the system to enhance search intensification at the earlier stage of the search. In addition, the users could succeed to design filters which could find 85% of their preferred fruit items.

Figure 4(c) and (d) show the output of examinees E_1 and E_3 who chose a lime and a tomato respectively. The filter designed by E_1 could succeed in finding the lime image areas, but failed to distinguish the papaya from background. The filter designed by E_3 to find the tomato founds all tomato image areas but missed the lime. Figure 5 shows the designed filter structure by E_3, and its transitions of $F(C)$, $Q(C)$, $P(C)$, and user-system cooperation changes. In Fig. 5(a), edges are categorized in two types for easy to understand; doted edges do not affect the output of the filter, and nodes connected by border edges are used to generate the output image. In Fig. 5(b), simple cooperation way transitions were not observed, and the role switch occurred at frequent intervals. This is because Non-IEC search required more time to produce good solutions than the time examinees expected.

5 Conclusions

This paper proposes a method for user-system cooperation in evolutionary computation aiming to optimize both quantitative and qualitative objective functions and its application system for image processing filter design. Experiments have shown that the proposed method allows users to design image filters based on their heuristics and preference. In future, we plan to utilize multi-objective optimization in order to produce various solutions.

Acknowledgement. This work was supported by Grant-in-Aid for Scientific Research (23700272) from the Ministry of Education, Culture, Sports, Science and Technology, Japan.

References

Amamiya, A., Miki, M., Hiroyasu, T.: Interactive genetic algorithm using initial individuals produced by support vector machine. In: The Science and Engineering Review of Doshisha University, vol. 50, no. 1, pp. 34–45 (2009)

Ando, D., Iba, H.: Interactive composition aid system by means of tree representation of musical phrase. In: Proceedings of the IEEE Congress on Evolutionary Computation, pp. 4258–4265 (2007)

Hitoyasu, T., Kobayashi, Y., Sasaki, Y., Tanaka, M., Miki, M., Yoshimi, M.: Discussion of evaluation methods for multiobjective interactive genetic algorithm. In: Proceedings of the World Automation Congress, CD-ROM (2010)

Horn, J.: The nature of niching: genetic algorithms and the evolution of optimal cooperative populations, p. 97008. Technical report IlliGAL Report No (1997)

Huang, W., Matsushita, D., Munemoto, J.: Interactive evolutionary computation (IEC) method of interior work (iw) design for use by non-design-professional Chinese residents. J. Asian Arch. Build. Eng. **5**(1), 91–98 (2006). http://ci.nii.ac.jp/naid/110004773753/en/

Kim, H.S., Cho, S.B.: Application of interactive genetic algorithm to fashion design. Eng. Appl. Artif. Intell. **13**(6), 635–644 (2000)

Maeda, H., Ono, S., Nakayama, S.: A fundamental study on the effectiveness of network-structured image filter generation method in traffic sign extraction. Technical report, IEICE on Pattern Recognition and Media Understanding, vol. 109, no. 470, pp. 383–388 (2010, in Japanese)

Maezono, M., Ono, S., Nakayama, S.: Automatic parameter tuning and bloat restriction in image processing filter generation using genetic programming. Trans. Jpn. Soc. Comput. Engineering and Science **2006**, 20060021 (2006)

Nagao, T., Masunanga, S.: Automatic construction of image transformation processes using genetic algorithm. In: Proceedings of International Conference on Image Processing, pp. 731–734 (1996)

Ono, S., Nakayama, S.: Fusion of interactive and non-interactive evolutionary computation for two-dimensional barcode decoration. In: Proceedings of the IEEE World Congress on Computational Intelligence (WCCI 2010), pp. 2570–2577 (2010)

Osaki, M., Takagi, H.: Reduction of the fatigue of human interactive ec operators: improvement of present interface by prediction of evaluation order. Jpn. Soc. Artif. Intell. **13**(5), 712–719 (1998). http://ci.nii.ac.jp/naid/110002808096/en/

Riesbeck, C.K., Schank, R.C.: Inside Case-Based Reasoning. Lawrence Erlbaum, Hillsdale (1989)

Satoh, H., Yamamura, M., Kobayashi, S.: Minimal generation gap model for gas considering both exploration and exploitation. In: Proceedings of the International Conference on Soft Computing, pp. 494–497 (1996)

Schank, R., Kassand, A., Riesbeck, C.: Inside Case-Based Explanation. Lawrence Erlbaum, Hillsdale (1994)

Shirakawa, S., Nagao, T.: Genetic image network (GIN): automatically construction of image processing algorithm. In: Proceedings of the International Workshop on Advance Image Technology (2007)

Takagi, H.: Interactive evolutionary computation - fusion of the capabilities of EC optimization and human evaluation. Proc. IEEE **89**, 1275–1296 (2001)

Unemi, T.: A design of multi-field user interface for simulated breeding. In: Proceedings of the Asian Fuzzy Systems Symposium, pp. 489–494 (1998)

Unemi, T., Nakada, E.: A tool for composing short music pieces by means of breeding. In: Proceedings of the IEEE International Conference Systems, Man and Cybernetics, pp. 3458–3463 (2001)

Metaheuristic Methods

Tabu Search Algorithm for the Vehicle Routing Problem with Time Windows, Heterogeneous Fleet, and Accessibility Restrictions

Joseph Sebastian Widagdo[(✉)] and Andi Cakravastia

Industrial Engineering Faculty of Industrial Technology, Bandung Institute of Technology, Jalan Ganeca 10, Bandung, West Java 40132, Indonesia
josephwidagdo@gmail.com, andi@mail.ti.itb.ac.id

Abstract. Hospitals or the other health service providers usually assign a third party service provider to manage their solid medical waste. In our case, it is known that the third party service's transportation cost is too expensive. Through the analysis, it is found that the main problem lies in determining solid medical waste pick-up routes.

In this research, a new method to determine pick-up routes will be proposed. The route proposed will minimize the company's transportation cost. In this case, the clients have their opening hours, the third party service providers has several kinds of vehicle, and there are accessibility restrictions for certain vehicle. The problem is known as VRPHETW with accessibility restrictions. The mathematical model and 2-phased tabu search will be used to solve this problem. The result shows that our proposed route is better than the existing route.

Keywords: Medical waste management · Transportation · VRPHETW
Accessibility restrictions · Transportation cost

1 Introduction

Health is an important aspect in human life. Without proper health care, people will get sick easily. In the long term, it also contributes to economic progress, as healthy populations work more productive and save more. Developed and emerging countries have started to improve their health care system. The increasing expenditures for both public and private sector of the U.S economy could be an evidence of this phenomenon (Nahata et al. 2005).

One of influencing factors for health performance is the ability to manage waste. Depending on the generation resource, wastes can be classified into household waste, industrial waste, and medical waste (Tabasi and Marthandan 2013). They also stated that medical waste is classified as one of the most dangerous waste. Therefore, medical waste management is important due to their potential as environmental hazards and their risks to human health (Mbarki et al. 2013).

B. K. Ane et al. (Eds.): WSC 2014, AISC 864, pp. 181–193, 2019.
https://doi.org/10.1007/978-3-030-00612-9_16

In recent years, concern over medical waste management has increased in Indonesia. Several years ago, government established regulations to standardize medical waste management. Ministry of Health was the person in charge for this regulation. Usually, health service providers manage its liquid medical waste using its Liquid Waste Treatment Plant. However, the providers use third party service to manage its solid medical waste.

In Java Island, Indonesia, the third parties have to travel more than 400 km daily to pick up medical waste from their clients. Each clients have opening hours and the third parties have different types of vehicle. In addition, there are accessibility restrictions for big vehicles. This problem is illustrated as can be seen in Fig. 1. Big vehicles are not allowed to serve clients with small access road. This paper is motivated by the high transportation cost faced by the third parties. Using their current system, their transportation cost is too expensive. Transportation cost depends on how its management determines its route and schedule in collecting medical waste from its clients.

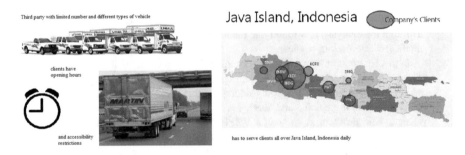

Fig. 1. Problem illustration

The problem is known as VRPHETW (*Vehicle Routing Problem with Heterogeneous Fleet and Time Windows*) with accessibility restrictions. In this research, a mathematical model and a tabu search algorithm will be proposed. A VPRHETW mathematical model proposed by Jiang et al. (2014) were used with some adjustments. These adjustments were made because there are accessibility restrictions in the real system. Meanwhile, in Jiang et al. (2014) accessibility restrictions are not accommodated.

The tabu search algorithm developed in this research is based on the 2-phased tabu search developed by Jiang et al. (2014). This 2-phased tabu search is inpired by holding list developed by Lau et al. (2003). Lau et al. (2003) introduced holding list in tabu search to improve the solution quality of the tabu search to solve the VRP with time windows and limited homogeneous vehicles. Our contribution is in proposing a mathematical model and a tabu search algorithm to solve this VRP problem. The result shows that our proposed route is better than the company's existing route and potentially reduces the company's transportation cost.

This paper is organized as follows. We review the literature on closely related topics in Sect. 2. Then, a formal description of the VRPHETW with accessibility restriction is given in Sect. 3. The outline and details of our proposed algorithm are explained in Sect. 4. After that, in Sect. 5 we provide computational results and analysis. Finally, in Sect. 6 we give conclusions.

2 Literature Review

VRP is a NP hard problem. It means that the time needed for analytical computation increases exponentially as the linear increase of problem complexity. Therefore, besides mathematical formulation, many authors also developed a heuristic method to solve the VRP problem.

The heterogeneous vehicle routing problem with time windows and accessibility restrictions (VRPHETW with accessibility restrictions) is defined as follows. There are limited number of vehicles of several types, the customers have their open hours, and there are vehicles that cannot serve certain customers due to certain reasons, for example the size of the road.

The vehicle routing problem with time windows (VRPTW) is one of the most studied extensions of the VRP. In this VRPTW, each customer has a time window $[e_i, l_i]$. Customers only want to be served in this time slot. Time windows are hard constraints in VRPTW. It means that it is possible to arrive to customer's location before its opening time, but the vehicle has to wait until the time windows open. However, many authors developed penalties in VRPTW with soft time windows. One of them is Badeau et al. (1997), who developed a VRPTW heuristic using a parallel tabu search heuristic. By using this heuristic, they claimed that parallelization of the original sequential algorithm does not reduce solution quality, for the same amount of computations, while providing substantial speed-ups in practice. Figliozzi (2010) proposed an iterative route construction and improvement algorithm for the VRP with soft time windows.

The original heterogeneous VRP is the unlimited heterogeneous VRP, usually called FSMVRP (Fleet Size and Mixed VRP). In this type of VRP, there is several types of vehicle and the number of vehicle is assumed to be unlimited. Renaud and Boctor (2002) presented a sweep-based algorithm for the unlimited heterogeneous VRP. The proposed algorithm first generates a large number of routes that are serviced by one or two vehicles. The selection of routes and vehicles to be used is then made by solving to optimality, in polynomial time, a set partitioning problem having a special structure. Lima et al. (2004) solved the unlimited heterogeneous VRP using a memetic algorithm or hybrid genetic algorithm. Choi and Tcha (2007) developed a column generation approach to the unlimited heterogeneous VRP. Liu et al. (2009) developed an effective genetic algorithm for the fleet size and mix vehicle routing problems. Prins (2009) presented two memetic algorithms (genetic algorithms hybridized with a local search) for the unlimited heterogeneous VRP. Brandao (2009) proposed a deterministic tabu search algorithm for the fleet size and mix vehicle routing problem.

The limited heterogeneous VRP (VRPHE) is a modified version of FSMVRP where the number of vehicle is assumed to be limited. Taillard (1999) developed a column generation method for the VRPHE. The method is claimed to be robust and efficient, particularly for medium and large size problem instances. Tarantilis et al. (2004) introduced a threshold accepting metaheuristic for the VRPHE. Li et al. (2007)

introduced a record-to-record travel algorithm for solving the heterogeneous fixed fleet vehicle routing problem. Imran et al. (2009) developed a variable neighborhood-based heuristic for the heterogeneous fleet VRP. Li et al. (2010) introduced an adaptive memory programming metaheuristic for the heterogeneous fixed fleet VRP. Brandao (2011) developed a tabu search algorithm for the limited heterogeneous VRP.

Many authors also developed VRP for more than one characteristics. Semet and Taillard (1993) developed a tabu search algorithm to solve VRP with limited heterogeneous vehicle, time windows, accessibility restriction. In addition, they also determine trailers' route. Brandao and Mercer (1997) developed a tabu search algorithm for the unlimited heterogeneous VRP. In this research, the vehicle can operate more than once in the planning horizon. They also considered an accessibility restriction and legal driving time per day and legal time breaks. Some of them included accessibility restrictions, drivers' break, and overload. Lau et al. (2003) developed a tabu search algorithm with holding list to solve the limited heterogeneous VRP with time windows (VRPHETW). Belfiore and Favero (2007) developed a mathematical model and a scatter search algorithm for heterogeneous unlimited VRP with time windows. Dell'Amico et al. (2007) also developed an algorithm for the fleet size and mix VRP with time windows. They developed a constructive insertion heuristic and a meta-heuristic algorithm. Paraskevopoulos et al. (2008) presented a two-phase solution framework based upon a hybridized tabu search, within a new reactive variable neighborhood search metaheuristic algorithm for solving the heterogeneous fixed fleet vehicle routing problem with time windows. Braysy et al. (2009) proposed a well-scalable metaheuristic for the unlimited heterogeneous VRP with time windows. The solution method combines the strengths of well-known threshold accepting and guided local search metaheuristics. Repoussis and Tarantilis (2010) developed an adaptive memory programming to solve the unlimited heterogeneous VRP with time windows. Kritikos and Ioannou (2013) also developed a mathematical model and an insertion heuristic for the unlimited heterogeneous VRP. In addition, they considered capacity as a soft constraint. Belfiore and Yoshizaki (2013) provided a mathematical formulation and an algorithm for the heterogeneous unlimited VRP with time windows and split deliveries. They developed a scatter search procedure to solve this problem. Gagliardi et al. (2014) introduced a first mathematical formulation and a heuristic algorithm for the VRP with pauses and time windows. This VRP applies well for long-haul transportation where drivers need some breaks within their work hours. Jiang et al. (2014) developed a two-phased tabu search algorithm with holding list adapted from Lau et al. (2003) to solve the VRPHETW.

3   Mathematical Formulation

The VRPHETW with accessibility restrictions can be defined on a graph $G(V,A)$. $V = \{0, 1, ..., n\}$ is the node set, where 0 represents the depot and n is the number of customers. $A = \{(i, j): 0 \leq i, j \leq n, i \neq j\}$ is the arc set. t_{ij} is travel time or distance that are associated with each arc (i, j). Each customers has a fixed demand d_i, open hours between e_i and l_i, service time associated with each costumer i s_i. Vehicle is represented by $k \in K$. Type of vehicle is represented by $c \in C$. If a vehicle k arrives at costumer i before its open hours at a_{ik} ($a_{ik} \leq e_i$), it has to wait for w_{ik} ($w_{ik} = e_i - a_{ik}$). There are c type of vehicle. Each c type of vehicle has its capacity q_c, there are only n_c vehicle available, latest returning time p_c, fixed cost f_c, and variable cost αc. The decisions variables are x_{ij}^k which is the decision to assign vehicle k to serve arc i and j, a_{ik} which is the arrival time of vehicle k in costumer i, and w_{ik} which is waiting time vehicle k in costumer i.

The accessibility restrictions in this research exists due to size of the road. Big vehicles cannot access customers that are located in small road. To accommodate this accessibility restrictions, we isolated big vehicles into set of big vehicles K_b. Also, we isolated small customers into set of small customers N_k. Vehicles that are included in set of K_b, cannot access customers that are included in set of N_k.

In this research, we use reference model from Jiang et al. (2014). However, there are slight differences between the real system and system in Jiang et al. (2014) that could be seen in Table 1.

Table 1. Comparison between reference model and developed model

Component	Reference model (Jiang et al. 2014)	Developed model
Performance criterion	Number-of –clients-served component is considered	All clients have to be served, so number of clients served become constraint
Performance criterion	Try to minimize number of vehicle used	There is no tendency to minimize number of vehicle used
Travelling time	There is no difference on each vehicle's travelling time	There is difference on each vehicle's travelling time
Accessibility restriction	There is no accessibility restriction	There is accessibility restriction

Therefore, these were the adjustments made in the research:

1. The number-of-clients-served component in objective function was not considered.
2. Big M in the fixed-and-variable-cost component in objective function was not considered.
3. Ratio parameter was added.
4. Constraint to accommodate accessibility restriction was added.

The VRPHETW with accessibility restrictions can be formulated as:

$$Min \sum_{c \in C} f_c \sum_{k \in S_c} \sum_{j \in N} x_{0j}^k + \sum_{c \in C} \alpha_c \sum_{k \in S_c} \sum_{(i,j) \in A} t_{ij} x_{ij}^k r_k \tag{1}$$

Subject to:

$$\sum_{j \in N_k} x_{ij}^k = 0 \quad \forall i \in V, \forall k \in K_b \tag{2}$$

$$\sum_{k\in K}\sum_{j\in V} x_{ij}^k = 1 \quad \forall i \in N \tag{3}$$

$$\sum_{k\in K}\sum_{i\in V} x_{ij}^k = 1 \quad \forall j \in N \tag{4}$$

$$\sum_{j\in N} x_{0j}^k \leq 1 \quad \forall k \in K \tag{5}$$

$$\sum_{i\in N} x_{i0}^k \leq 1 \quad \forall k \in K \tag{6}$$

$$\sum_{i\in V} x_{ij}^k = \sum_{i\in V} x_{ji}^k \quad \forall j \in V, k \in K \tag{7}$$

$$\sum_{i\in N} d_i \sum_{j\in \Delta_{ij}^+} x_{ij}^k \leq q_c \quad \forall k \in S_c, c \in C \tag{8}$$

$$x_{ij}^k \left(a_{ik} + w_{ik} + s_i + t_{ij} - a_{jk} \right) = 0 \quad \forall k \in K, (i,j) \in A \tag{9}$$

$$a_{ik} \leq l_i \sum_{j\in \Delta_i^+} x_{ij}^k \quad \forall k \in K, i \in N \tag{10}$$

$$e_i \sum_{j\in \Delta_i^+} x_{ij}^k \leq a_{ik} + w_{ik} \leq l_i \sum_{j\in \Delta_i^+} x_{ij}^k \quad \forall k \in K, i \in N \tag{11}$$

$$E \leq a_{0k} \leq L \quad \forall k \in K \tag{12}$$

$$a_{0k} \leq p_c \quad \forall k \in S_c, c \in C \tag{13}$$

$$\sum_{k\in S_c}\sum_{j\in N} x_{0j}^k \leq n_c \quad \forall c \in C \tag{14}$$

$$w_{ik} \geq 0 \quad \forall k \in K, i \in N \tag{15}$$

$$a_{ik} \geq 0 \quad \forall k \in K, i \in N \tag{16}$$

$$x_{ij}^k \in \{0,1\} \quad \forall k \in K, (i,j) \in A \tag{17}$$

The objective function (1) consists of fixed cost component and variable cost component. Constraint (2) accommodates the accessibility restrictions. Constraint (3) and (4) ensures that every customer has to be visited exactly once. Constraint (5) and (6) ensures that vehicles will not be used more than once in the same planning horizon. Constraint (7) states that every vehicle entering a customer has to exit that customer. Constraint (8) is a capacity constraint. Constraint (9) defines the sequence of events at every customer i. Constraint (10) ensures that vehicles cannot serve customer after its open hours. Constraint (11) to (13) is a time windows constraint. Constraint (14) is a limited number of vehicle constraint. Last, constraint (15) to (17) restrict the values of the variables.

4 Proposed Algorithm

In this research, we use a construction heuristic sequential insertion and an improvement metaheuristic two-phased tabu search algorithm proposed by Jiang et al. (2014). In this algorithm, Jiang et al. (2014) adapted holding list from Lau et al. (2003). Holding list is a phantom route such that insertion of a customer to the holding list is always feasible without any cost. However, there were adjustments made to accommodate the accessibility restrictions.

Pre-processing Data
Before starting the tabu search algorithm, we sort the customers and the vehicles. Jiang et al. (2014) proposed four rules in sorting the customers:

1. Earliest ready time rule: customer with the earliest ready time will be served first.
2. Tightest window rule: customer with the tightest open hours will be served first.
3. Greatest distance rule: customer with the farthest distance from the depot will be served first.
4. Greatest demand rule: customer with the biggest demand will be served first.

In this research, we use the greatest distance rule. We use this rule because we would like to minimize the total distance traveled by the vehicles.

Jiang et al. (2014) also proposed three rules in sorting the vehicle:

1. Greatest capacity rule: vehicle with the greatest capacity will be used first.
2. Smallest capacity rule: vehicle with the smallest capacity will be used first.
3. Greedy insertion rule: a simple greedy insertion is used to choose which type of vehicle will be used if only one type of vehicle can be used.

In this research, we use the greatest capacity rule. We use this rule because in the objective function, there is a fixed cost component.

Phase 1
The aim of this phase is to place each customer in the best location and route possible. 4 moves that are used in this phase, they are:

1. Transfer: Movement of client p from route r to route s location q.
2. Exchange: Exchange of client p from route r with client at route s location q.
3. Transfer from holding list: Movement of client p from holding list to route r location q.
4. Exchange with holding list: Exchange of client p from holding list with client at route r location q.

Phase 2

The purpose of this phase is to move customers from bigger vehicle as many as possible to the smaller ones. This move stands on assumption that the bigger the vehicle, the more expensive the operating cost. There will be only one move in this phase. Multitransfer move will be used in this phase. Multitransfer is defined as movement of client p from route ar1 location g, g + 1,..., until the last position, to last position of route ar2.

Two-phased Tabu Search Algorithm

As stated before, adjustments had to be made in the algorithm. Differences and adjustments made are explained below:

1. In Jiang et al. (2014), algorithm can move to the 2nd phase only if there is no client left unserved or all of the vehicles have been used. In this research, both of the requirements have to be met.
2. In the last sequential insertion, every vehicle is considered.

The two-phased tabu search algorithm can be described as follows:

Set numVeh, StepSize, CountLimit1, CountLimit2, CountLimitS

        Phase1 :
        Until $|H| = 0$ and numVeh $= |K|$
          If numVeh $= |K|$
              Do Sequential Insertion for StepSize only
          Else
              Do Sequential Insertion for all $|K|$ vehicle
          Endif
          Set Count $= 0$
            While Count $\leq$ CountLimit1
            Do TS based on numVeh
            If better solution found
              Set Count $= 0$
            Else
              Set Count $=$ Count $+ 1$
            Endif
            EndWhile
        Set numVeh $= \min($numVeh $+$ StepSize, $|K|)$
        End Until

Phase 2 :
Set CountS = 0
While CountS ≤ CountLimitS
    Do TSS
    Set Count2 = 0
    While Count2 ≤ CountLimit2
      Do TS
        If better solution found
            Set Count2 = 0
        Else
            Set Count2 = Count2 + 1
        Endif
      EndWhile
    If better solution found
      Set CountS = 0
    Else
      Set CountS= CountS + 1
    Endif
    EndWhile

After pre-processing data, we proceed to phase 1. In this phase, we consider only *numVeh* vehicles. *numVeh* denotes the number of vehicles used in a certain iteration. Then, based on that number of vehicle, we use Sequential Insertion to construct an initial solution. If the number of vehicle equals to number of vehicle available |K|, Sequential Insertion to all of the vehicle is used. After that, TS is used. TS represents one iteration of tabu search procedure with 4 moves described in phase 1. If solution found is not better than the best solution known so far, then *count* is added by 1. In other words, this counter counts non-improving moves. This loop continues as long as *count* doesn't exceed *countlimit1*. Then, we add the number of vehicle by *Stepsize* and go back to TS procedure based on *numVeh*. phase 1 ends whenever there is no clients left unserved and *numVeh* equals number of vehicle provided.

In phase 2, there are two counters, *count2* and *countS*. *Count2* represents the counter for TS procedure, while *countS* represents the counter for TSS procedure. TSS represents on iteration of post-processing tabu search which move is described before. First, we do a TSS procedure to move as many clients from bigger vehicle to the smaller one as possible. Then, a TS procedure is done to improve the solution. This phase ends when *countS* exceed *countlimitS*.

5   Computational Results and Analysis

The computation is done using data that can be seen in attachment A. Table 2 shows algorithm's parameter used in the computation. These values were chosen because they showed best objective function among the other combinations.

Table 2. Values of algorithm's parameter

Parameter	Values
countLimit1	3
countLimit2	7
countLimitS	7
maxNeigbour	500
StepSize	3
Service time	15
Departure time	180
numVeh	3
Tabulength	30

In this algorithm, departure time is defined as a parameter, at the 180th minute, that is 3 a.m. Tables 3 and 4 shows the differences between the existing route and the proposed route. From Table 4, we can see that the number of wheeled bin doesn't exceed the vehicle's capacity. The solution also meets the accessibility restrictions and the time windows.

Table 3. Existing route

EXISTING

Fleet	Sequence																				Wheeled Bin	Cost	
Fleet 1	0	45	46	47	48	49	50	51	52	0											20.5	Rp	656,810
Fleet 2	0	1	2	3	4	5	6	7	8	9	10	11	12	0							19.5	Rp	682,063
Fleet 3	0	13	14	15	16	17	18	19	20	21	0										18.5	Rp	980,143
Fleet 4	0	38	39	40	41	42	43	44	0												20.5	Rp	622,095
Fleet 5	0	56	57	58	59	0															21	Rp	997,781
Fleet 6	0	22	23	24	25	0															16	Rp	961,080
Fleet 7	0	26	27	28	29	30	31	32	33	0											13	Rp	755,628
Fleet 8	0	78	79	80	0																14	Rp	481,724
Fleet 9	0	34	35	36	37	0															6	Rp	462,018
Fleet 10	0	53	54	55	81	82	83	0													9	Rp	499,044
Fleet 11	0	60	61	62	63	64	65	66	67	68	69	70	71	72	73	74	75	76	77	0	8.65	Rp	1,055,938
Total																					166.65	Rp	815,324

The result shows that using this developed tabu search, new routes can be proposed to the third party. The route proposed potentially reduces the company's transportation cost by Rp 39.231.000 annually. The movement of the objective function for each iteration can be seen in Fig. 1. A good algorithm has to show convergence result. In this research, convergence is guaranteed by using the pre-processing data. By using the greatest distance rule, the initial solution made using sequential insertion will produce

Table 4. Proposed route

PROPOSED

Fleet	Sequence																			Wheeled Bin	Cost	
Fleet 1	0	55	80	34	79	78	81	43	0											24	Rp	737,627
Fleet 2	0	2	4	6	5	48	47	46	0											17.5	Rp	550,662
Fleet 3	0	42	45	38	39	40	41	50	49	0										18	Rp	592,992
Fleet 4	0	1	3	7	8	9	51	52	10	11	12	0								20	Rp	682,063
Fleet 5	0	30	21	29	22	13	14	17	0											20.5	Rp	1,012,773
Fleet 6	0	44	20	15	18	19	16	59	58	0										18	Rp	908,225
Fleet 7	0	57	56	26	32	33	31	28	27	0										18	Rp	1,000,295
Fleet 8	0	24	25	23	0															13	Rp	875,880
Fleet 9	0	53	82	54	83	35	36	37	0											9	Rp	577,629
Fleet 10	0	61	60	63	73	74	75	77	62	64	69	71	72	65	68	70	76	66	67	0	8.65	Rp 1,085,408
Fleet 11																						
Total																				166.65	Rp	8,023,554
Time																					422.9 seconds	

better result because the relative distance between customers is minimized. Then, by considering the biggest vehicle in the first attempt using the biggest capacity rule, we could minimize the usage of vehicle. Thus, the number of the vehicle used will be minimized, and so will the objective function (Fig. 2).

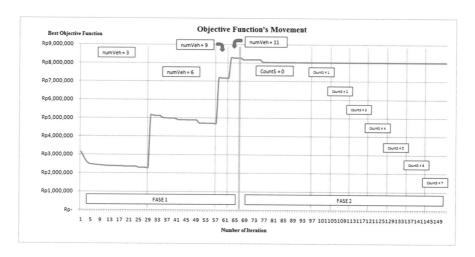

Fig. 2. Objective function's movement each iteration

Table 5 shows the result of sensitivity analysis. The sensitivity analysis is done by observing the change in the objective function due to change in the both mathematical model parameters and the algorithm parameters.

Table 5. Sensitivity analysis result

Model's parameter	Result	Algorithm's parameter	Result
Demand	Sensitif	CountLimit1	Insensitive
Capacity	Sensitive	CountLimit2	Insensitive
Service time	Insensitive	CountLimitS	Insensitive
Travel time	Sensitive	maxNeighbor	Sensitive
Opening hour	Insensitive	StepSize	Insensitive
Velocity ratio	Insensitive	NumVeh	Sensitive
Fixed cost	Insensitive	*Tabu length*	Insensitive
Variable cost	Insensitive		

6   Conclusions

This research leads to the following conclusions:

1. This research proposes new routes using the 2-phased tabu search algorithm.
2. Adjustments were made in order to accommodate accessibility restrictions.
3. The third party could potentially reduce its transportation cost by Rp 39.231.000 annualy.
4. Using this method, the third parties can determine their route quickly without extra effort from its employees.

References

Badeau, P., Guertin, F., Gendreau, Michel., Jean-Yves, P., Taillard, E.: A parrallel tabu search heuristic for the vehicle routing problem with time windows. Transp. Res.-C **5**(2), 109–122 (1997)

Belfiore, P., Favero, L.P.L.: Scatter search for the fleet size and mix vehicle routing problem with time windows. CEJOR **15**, 351–368 (2007)

Belfiore, P., Yoshizaki, H.T.Y.: Heuristic methods for the fleet size and mix vehicle routing problem with time windows and split deliveries. Comput. Ind. Eng. **64**, 589–601 (2013)

Brandao, J.: A deterministic tabu search algorithm for the fleet size and mix vehicle routing problem. Eur. J. Oper. Res. **195**, 716–728 (2009)

Brandao, J.: A tabu search algorithm for the heterogeneous fixed fleet vehicle routing problem. Comput. Oper. Res. **38**, 140–151 (2011)

Brandao, J., Mercer, A.: A tabu search algorithm for the multi-trip vehicle routing and scheduling problem. Eur. J. Oper. Res. **100**, 180–191 (1997)

Braysy, O., Porkka, P.P., Dullaert, W., Repoussis, P.P., Tarantilis, C.D.: A well scalable metaheuristic for the fleet size and mix vehicle routing problem with time windows. Expert Syst. Appl. **36**, 8460–8475 (2009)

Choi, E., Tcha, D.-W.: A column generation approach to the heterogeneous fleet vehicle routing problem. Comput. Oper. Res. **34**, 2080–2095 (2007)

Dell'Amico, M., Corrado, M.P., Daniele, V.: Heuristic approaches for the fleet size and mix vehicle routing problem with time windows. Transp. Sci. **41**(4), 516–526 (2007)

Figliozzi, M.A.: An iterative route construction and improvement algorithm for the vehicle routing problem with soft time windows. Transp. Res. Part C: Emerg. Technol. **18**(5), 668–679 (2010)

Gagliardi, J.-P., Renaud, J., Ruiz, A., Coelho, L.C.: The vehicle routing problem with pauses. CIRRELT, May 2014

Imran, A., Salhi, S., Wassan, N.A.: A variable neighborhood-based heuristic for the heterogeneous fleet vehicle routing problem. Eur. J. Oper. Res. **197**, 509–518 (2009)

Jiang, J., Ng, K.M., Poh, K.L., Teo, K.M.: Vehicle routing problem with a heterogeneous fleet and time windows. Expert Syst. Appl. **41**, 3748–3760 (2014)

Kritikos, M.N., George, I.: The heterogeneous fleet vehicle routing problem with overloads and time windows. Int. J. Prod. Econ. **144**, 68–75 (2013)

Lau, H.C., Sim, M., Teo, K.M.: Vehicle routing problem with time windows and a limited number of vehicles. Eur. J. Oper. Res. **148**, 559–569 (2003)

Li, F., Golden, B., Wasil, E.: A record-to-record travel algorithm for solving the heterogeneous fleet vehicle routing problem. Comput. Oper. Res. **34**, 2734–2742 (2007)

Li, X., Tian, P., Aneja, Y.P.: An adaptive memory programming metaheuristic for the heterogeneous fixed fleet vehicle routing problem. Transp. Res. Part E **46**, 1111–1127 (2010)

Lima, C.M.R.R., Goldbarg, M.C., Goldbarg, E.F.G.: A memetic algorithm for the heterogeneous fleet vehicle routing problem. Electorn. Notes Discret. Math. **18**, 171–176 (2004)

Liu, S., Huang, W., Ma, H.: An effective genetic algorithm for the fleet size and mix vehicle routing problems. Transp. Res. Part E **45**, 434–445 (2009)

Mbarki, A., Kabbachi, B., Ezaidi, A., Benssaou, M.: Medical waste management: a case study of the souss-massa-draa region, Morocco. J. Environ. Prot. **4**, 914–919 (2013)

Nahata, B., Ostaszewski, K., Sahoo, P.: Rising health care expenditures: a demand-side analysis. J. Insur. Issues **28**(1), 88–102 (2005)

Paraskevopoulus, D.C., Repoussis, P.P., Tarantilis, C.D., Ioannou, G., Prastacos, G.P.: A reactive variable neighborhood tabu search for the heterogeneous fleet vehicle routing problem with time windows. J. Heuristics **14**, 425–455 (2008)

Prins, C.: Two memetic algorithms for heterogeneous fleet vehicle routing problems. Eng. Appl. Artif. Intell. **22**, 916–928 (2009)

Renaud, J., Boctor, F.F.: A sweep-based algorithm for the fleet size and mix vehicle routing problem. Eur. J. Oper. Res. **140**, 618–628 (2002)

Repoussis, P.P., Tarantilis, C.D.: Solving the fleet size and mix vehicle routing problem with time windows via adaptive memory programming. Transp. Res. Part C **18**, 695–712 (2010)

Semet, F., Taillard, E.: Solving real-life vehicle routing problems efficiently using tabu search. Ann. Oper. Res. **41**, 469–488 (1993)

Tabasi, R., Marthandan, G.: Clinical waste management: a review on important factors in clinical waste generation rate. Int. J. Sci. Technol. **3**(3) (2013)

Taillard, E.D.: A heuristic column generation method for the heterogeneous fleet VRP. Oper. Res. **33**, 1–14 (1999)

Tarantilis, C.D., Kiranoudis, C.T., Vassiliadis, V.S.: A threshold accepting metaheuristic for the heterogeneous fixed fleet vehicle routing problem. Eur. J. Oper. Res. **152**, 148–158 (2004)

Applying Multi-objective Particle Swarm Optimization for Solving Capacitated Vehicle Routing Problem with Load Balancing

Dominico Laksma Paramestha, The Jin Ai$^{(\boxtimes)}$,
and Slamet Setio Wigati

Industrial Engineering Department, Universitas Atma Jaya Yogyakarta,
Yogyakarta, Indonesia
dlaksma@live.com,
{the.jinai,slamet.wigati}@uajy.ac.id

Abstract. This research considers workload balance in the Capacitated Vehicle Routing Problem (CVRP), which can be called the CVRP with Load Balancing (CVRPLB). In addition to find a fleet of distribution vehicles with minimum cost, the workload balance among vehicles are considered. To solve the CVRPLB, a Multi-Objective Particle Swarm Optimization algorithm is applied and implemented. The performance of the algorithm is tested on the Christofides et al. [1] test instances.

1 Introduction

Vehicle Routing Problem (VRP) is a typical distribution cases in which there are some customers to be served by a number of vehicles on a route that begins and ends at a depot. Route of vehicles are created so that they satisfy some operational constraints and the total travel cost is minimized [1, 2]. In other words, the VRP is used to design the optimal route for a number of vehicles to serve multiple customers with a set of constraints attached to the problem [3].

The simplest VRP variant is called the Capacitated Vehicle Routing Problem (CVRP), in which a number of identical vehicles starting from a single depot need to serve some customers. Due to its importance in the field of physical distribution and logistics, the CVRP has been attracted many researchers for more than five decades [4]. Various solution methodologies are available now for solving the CVRP, including some exact algorithms [5–7], classical heuristics [8–10], and several metaheuristics methods, such as genetic algorithm [11, 12], tabu search [13, 14], ant colony optimization [15, 16], and particle swarm optimization [17, 18].

Based on CVRP definition, all the solution methodologies proposed for solving the CVRP is focus on finding routes with the lowest transportation costs, which is indicated by the smallest number of vehicles and the shortest travel distance. As results, the solution does not consider the workload balance among vehicles, i.e. the load of vehicle to be delivered to the customers for satisfying demand. Table 1 shows the best known solution of CVRP benchmark data [1] that exist today. If the load imbalance is

© Springer Nature Switzerland AG 2019
B. K. Ane et al. (Eds.): WSC 2014, AISC 864, pp. 194–204, 2019.
https://doi.org/10.1007/978-3-030-00612-9_17

defined as the range of the vehicle load, i.e. the difference between the highest and the smallest vehicle load, it appears that there is a very high load imbalance experienced by the best known solution.

Table 1. Best known solution of CVRP benchmark data

Problem	Min	Max	Average	Range	% Range to average
VRPNC1	149	160	155.4	11	7.1%
VRPNC2	126	140	136.4	14	10.3%
VRPNC3	108	199	182.3	91	49.9%
VRPNC4	64	200	186.3	136	73.0%
VRPNC5	19	200	187.4	181	96.6%
VRPNC6	80	155	129.5	75	57.9%
VRPNC7	87	140	123.5	53	42.9%
VRPNC8	93	199	162.0	106	65.4%
VRPNC9	109	198	159.6	89	55.7%
VRPNC10	54	200	177.0	146	82.5%
VRPNC11	188	200	196.4	12	6.1%
VRPNC12	150	200	181.0	50	27.6%
VRPNC13	92	162	125.0	70	56.0%
VRPNC14	60	200	164.5	140	85.1%

In the operational level, there is a highly potential conflict due to load imbalance among vehicle operators since often the performance measurement and remuneration system of operators take the load being transported into account. Indirectly, this conflict will degrade the performance of the company. In other words, since the employees are important factor to the sustainability of the companies and the competitiveness of companies is affected by how the companies treat their employees in a fair manner, load balancing is one effort that can be done to treat employees fairly [19]. Thus, there is a need to develop a methodology to obtain vehicle routing solutions that also make the vehicle load balanced and in addition to obtain the lowest transportation costs. In this paper, we consider a problem called the Capacitated Vehicle Routing Problem with Load Balancing (CVRPLB). The CVRPLB is CVRP with two objectives, which are minimizing total transportation cost and minimizing the range of vehicle load. After the CVRPLB was proposed and solved using a heuristic algorithm [19], numerous researchers have been working with several problems related to the CVRPLB as affirmation of the importance of balancing issue in the VRP area. A very similar problem with CVRPLB called balanced cargo vehicle routing problem with time windows have been proposed with objectives to minimize cost and cargo load imbalance. The problem was solved by an approach based on the free disposal hull (FDH) method of data envelopment analysis [20]. Another problem called skill vehicle routing problem, in which considers skill requirements given on demand nodes, has been proposed in order to minimize cost and load imbalance. This problem was solved using two steps mathematical programming approach [21]. A combination of vehicle routing and two-dimensional packing problem has been addressed with criteria of cost and load imbalance in term of the loading difference between the most and the least loaded route. A non-dominated sorting genetic algorithm-II (NSGA-II) has been applied for solving the problem [22]. An uncapacitated stochastic vehicle routing problem, in which vehicle depot locations are fixed, and client locations in a service region are following a given probability density function, has been proposed. An algorithm for partitioning the service region while balancing the workloads of all vehicles has also been proposed [23]. A vehicle routing problems with time windows has been proposed with three objectives, that are total traveled distance, distance imbalance, and load imbalance. Two approaches have been proposed to solve this problem, which are a multi-start multi-objective evolutionary algorithm with simulated annealing [24] and multiple temperature pareto simulated annealing [25]. Another problem called vehicle routing problem with route balancing has been proposed in order to minimize total traveling cost and route imbalance in term of range of traveled distance, which is the difference between the longest and shortest route. Some techniques have been proposed to tackle this problem such as a hybrid heuristics algorithm called multi-start split based path relinking (MSSPR) [26] and greedy randomized adaptive search procedure with advanced starting point (GRASP-ASP) [27]. Another VRP related problem called location routing problem has been proposed with two objectives that are of cost and route imbalance in term of range of traveled distance. Two metaheuristic solution algorithms based on the scatter tabu search procedure for non-linear multi-objective optimization (SSPMO) and on the NSGA-II have been proposed for solving this problem [28].

This paper treats the CVRPLB as a Multi-Objective Optimization (MOO) and tried to solve it using Multi-Objective Particle Swarm Optimization (MOPSO). As reviewed above, PSO is never being used for solving the CVRPLB and its related problem. Hence, the capability of PSO can be evaluated here in order to tackle the CVRPLB. Moreover, the application of PSO for solving the CVRPLB can be considered as a foundation for tackling another related problems since usually an effective techniques for solving the CVRP can be easily applied for solving more complex variants of VRP.

2   CVRPLB Definition

Following the classical definition of the CVRP, the CVRPLB can be formally defined as follows. Let $G = (V, A)$ be a graph where $V = \{v_0, v_1, \ldots, v_n\}$ is a vertex set, and $A = \{(v_i, v_j) | v_i, v_j \in V, i \neq j\}$ is an arc set. Vertex v_0 represents a depot, while the remaining vertices correspond to customers. Associated with A are a cost or distance matrix (d_{ij}) and a travel time matrix (t_{ij}). Each customer has a non-negative demand q_i and a service time s_i. A fleet of m identical vehicles of capacity Q is based at the depot. The CVRPLB consists of designing a set of m routes such that

(1)   each route starts and ends at the depot,
(2)   each customer is visited exactly once by exactly one vehicle,
(3)   the total demand of each route does not exceed Q,

(4)   the total duration of each route including travel and service times does not exceed a preset limit D,

(5)   the total transportation cost and the range of vehicle load are minimized.

3   MOPSO Framework for Solving CVRPLB

Particle Swarm Optimization (PSO) is a population-based searching method imitating the behavior of swarm organism. For a single objective optimization problem, an instance of decision variables is represented by a particle in which its position has multi dimensions. A swarm of particles, which represents population of solutions, is moving during iteration step procedure, which represents evaluation of different population of solutions. The movement of particles are driven by movement equations that show the behavior of swarm organism: the cognitive and social behavior. The movement equations are updating two properties of particle, which are velocity and position. A PSO variant called GLNPSO that has three different social behavior terms called global best, local best, and nearest neighbor best with its corresponding acceleration constant (c_g, c_l, and c_n), i.e. the movement of particle in this variant is following these equations:

$$\omega_{lh}(\tau + 1) = w(\tau)\omega_{lh}(\tau) + c_p u \left(\psi_{lh} - \theta_{lh}(\tau)\right) + c_g u \left(\psi_{gh} - \theta_{lh}(\tau)\right) + c_l u \left(\psi_{lh}^L - \theta_{lh}(\tau)\right)$$
$$+ c_n u \left(\psi_{lh}^N - \theta_{lh}(\tau)\right)$$
(1)

$$\theta_{lh}(\tau + 1) = \theta_{lh}(\tau) + \omega_{lh}(\tau + 1)$$
(2)

where τ is iteration index, l is particle index, h is dimension index, u is uniform random number in interval [0,1], $w(\tau)$ is inertia weight in the iteration τ, ω_{lh} is velocity of particle l at the dimension h in the iteration τ, θ_{lh} is position of particle l at the dimension h in the iteration τ, ψ_{lh} is personal best position (pbest) of particle l at the dimension h, ψ_{gh} is global best position (gbest) at the dimension h, ψ_{lh}^L is local best position of particle l at the dimension h, and ψ_{lh}^N is nearest neighbor best position of particle l at the dimension h [20].

As mentioned before, PSO application for solving CVRP has been successfully demonstrated based on the GLNPSO [18]. This research is applying solution representation SR-1 presented in [18], that is $(m + 2n)$-dimension particle for representing CVRP problem with m customers and n vehicles, with corresponding decoding method to translate an instance of particle position into a typical of vehicles route.

In general, the Multi Objective Particle Swarm Optimization (MOPSO) Framework for solving CVRPB is proposed in Fig. 1. The MOPSO is searching for the Pareto optimal front, which consists of a set of non-dominated solutions, instead of a single best solution. Therefore, the MOPSO mechanism is different with the single objective PSO in terms of storage of elite group or non-dominated solutions, selection of a reference particle or guidance to drive the particles movement, and definition of movement strategy, i.e. how to use the reference particles as search guidance.

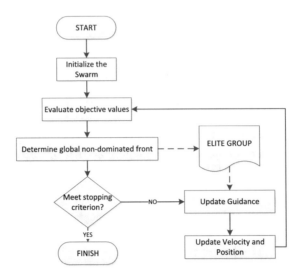

Fig. 1. MOPSO Framework

Following the methodology proposed in [20], the elite group is sorted based on non dominated sort presented in NSGA-II algorithm. Also, there are six possibilities of movement strategies to implement for replacing the global term in original PSO movement equation, i.e. $c_g u\left(\psi_{gh} - \theta_{lh}(\tau)\right)$ inside Eq. (1), which are called MS1, MS2, MS3, MS4, MS5, MS6. In MS1, global guidance (g) is randomly selected among elite group which is located in the least crowded area. Therefore, the global term is similar with original PSO, but the particle index g is changed. In MS2, a perturbation based on Differential Evolution concept is performed. In this movement strategy, the global term is replaced by $c_g u\left(\psi_{R1,h} - \psi_{R2,h}\right)$, in which the $R1$ and $R2$ are indices of two elite group members that are selected from top $t\%$ of most crowded area and top $b\%$ of least crowded area, respectively. MS3 is a strategy for exploring the unexplored space in the non-dominated front. At first, the members of the elite group are sorted ascending based on the objective function values. After that, unexplored space is identified by calculating the objective function gap between pairs of consecutive members. If a gap of pair members is greater than x% of the range of objective function values in the elite group, the pair is stored as the boundary of an unexplored space. During particle movement, an unexplored space is randomly selected. If E_1 and E_2 are the boundary of the selected unexplored space in the non-dominated front, the global term is replaced by $c_g u[(E_{1h} - \theta_{lh}(\tau)) + r(E_{1h} - E_{2h})]$. The MS4 is combination of MS1 and MS2. The MS5 is using mixed group of particles to explore solution space, each of it using original PSO movement, MS1, MS2, and MS3, respectively. Meanwhile, MS6 is using $n + 1$ group of particles or sub-swarms in which n is the number objective functions. The first n sub-swarms will search for the optimal solution corresponding to each objective function just like the tradition PSO and the last sub-swarm will minimize the adaptive weighted function of objective function.

In the MOPSO Framework for CVRPLB, the updating of particles' position is following Eq. (2). Once a particle position is updated, a corresponding set of vehicle routes can be generated following the decoding method of SR-1 [18]. Once a set of vehicle routes is generated, its corresponding objective function values can be calculated, i.e. its total transportation cost and the range of vehicle load. Based on both objective function values of several set of vehicle routes, the non-dominated front can be updated. These steps are repeated over some iterations, and the final non-dominated solutions or the Pareto optimal front can be determined at the end of iteration process.

4   Computational Demonstrations

The MOPSO Framework presented in Sect. 3 above is being implemented using C# programming language. A C# programming library called ET-Lib is being used to support the implementation of the framework, since it contains the library for the multi objective PSO [29]. For computational demonstrations, 14 CVRP instances called VRPNC [1] are selected. The VRPNC data is a well known CVRP data, in which each instance describes a particular CVRP data such as of number of customers, coordinate of the customers, coordinate of the depot, customers demand, vehicles capacity, and maximum route time.

The computational demonstrations are performed in a computer with Intel Core i3 2.1 GHz processor and 2 GB of RAM. The PSO parameter being used in the experiment are number of iteration = 500, number of particle = 50, initial $w = 0.9$, final $w = 0.4$, $c_p = 1$, $c_g = 1$, $c_l = 1$, $c_n = 1$, and number of replication = 1. Initially, the number of used vehicle is equal to the one used in best known solution. However, number of vehicle may be increased one by one, in the case of any customer cannot being served by any vehicle. The results of the proposed framework using MS1 are presented in Table 2. In this table, the Pareto front obtained is presented in the list of first (total cost) and second (range of vehicle load) objectives. For comparison purposes, the running time of the framework is presented as TimeCals (in second) and the objectives of its corresponding best known solution (BKS) are also presented in the last column.

The proposed framework is able to generate some alternatives solutions for case of VRPNC1- VRPNC5, with smaller range of vehicle load (the second objective) compared to its corresponding best known solution (BKS). It is implied that the solutions generated by this framework are resulted in a better workload balance compared to its corresponding BKS. It is also shown that the reduction of the range of vehicle load resulted in relatively low additional of cost (the first objective).

Different result showed in the experiment of using VRPNC6-VRPNC10 since these cases have constraints on the maximum distance for every generated route. This constraint presence resulted in a lot of solutions that cannot serve all customers, whenever the number of vehicles is similar with its corresponding best known solution. Therefore, experiments were done by adding the number of vehicles one by one, starting from the number of vehicles in its BKS, until the framework is able to generate majority of solutions which could serve all customers. Table 2 also summarizes the results from this experiment. Similarly to the previous result, the total cost is increasing

Table 2. Computational results of VRPNC1-VRPNC14

Instance	Computational results	BKS
VRPNC1	(537.26, 8); (537.81, 5); (538.11, 3); (562.61, 2); (617.78, 1) Vehicle: 5; TimeCalc: 16.1 s	(524.61, 11) Vehicle: 5
VRPNC2	(902.89, 7); (911.73, 6); (962.67, 5); (988.58, 4); (1090.32, 3) Vehicle: 10; TimeCalc: 26.5 s	(835.26, 14) Vehicle: 10
VRPNC3	(873.31, 43); (878.31, 37); (882.07, 33); (883.33, 16) (889.63, 10); (891.58, 8) Vehicle: 8; TimeCalc: 40.8 s	(826.14, 91) Vehicle: 8
VRPNC4	(1097.11, 29); (1102.09, 28); (1102.42, 27); (1103.78, 26); (1107.41, 25); (1107.62, 23); (1109.57, 22); (1109.83, 19); (1111.18, 18); (1233.24, 17) Vehicle: 12; TimeCalc: 69.3 s	(1028.4, 136) Vehicle: 12
VRPNC5	(1413.47, 30); (1413.56, 29); (1414.78, 27); (1418.25, 25); (1424.22, 24); (1436.83, 22); (1461.40, 20); (1544.39, 18) Vehicle: 17; TimeCalc: 105.7 s	(1291.30, 181) Vehicle: 17
VRPNC6	(572.48, 44); (573.47, 40); (573.77, 31); (574.08, 25); (575.32, 12); (581.51, 7); (593.60, 6) Vehicle: 6; TimeCalc: 17.7 s	(555.43, 75) Vehicle: 6
VRPNC7	(1126.17, 122); (1129.57, 111); (1131.44, 95); (1137.16, 75); (1147.87, 69); (1150.18, 64); (1159.32, 62); (1160.25, 52); (1163.75, 44); (1166.99, 44); (1178.76, 40); (1182.00, 36); (1183.10, 34); (1286.13, 32); Vehicle: 16; TimeCalc: 35.2 s	(909.68, 53) Vehicle: 11
VRPNC8	(1012.43, 116); (1012.84, 65); (1015.28, 51); (1018.16, 46); (1030.91, 45); (1039.77, 35); (1050.70, 34); (1063.04, 30); (1114.62, 26); (1133.76, 24) Vehicle: 11; TimeCalc: 47.3 s	(865.94, 106) Vehicle: 9
VRPNC9	(1351.91, 76); (1357.01, 67); (1360.26, 64); (1360.49, 62); (1374.44, 47); (1377.39, 46); (1377.39, 41); (1380.25, 40); (1585.50, 38) Vehicle: 17; TimeCalc: 84.4 s	(1162.55, 89) Vehicle: 14
VRPNC10	(1667.01, 127); (1668.93, 97); (1673.84, 88); (1681.26, 81); (1681.45, 76); (1682.85, 70); (1693.29, 70); (1697.53, 66); (1704.19, 64); (1706.49, 63); (1729.35, 60); (1737.30, 57); (1745.47, 54); Vehicle: 22; TimeCalc: 134.5 s	(1291.30, 181) Vehicle: 18
VRPNC11	(1063.99, 8); (1071.07, 7); (1072.01, 6); (1073.97, 5) (1104.04, 4); (1108.08, 3); (1112.69, 2) Vehicle: 7; TimeCalc: 59.2 s	(1042.10, 12) Vehicle: 5
VRPNC12	(821.96, 50); (822.65, 40); (838.03, 30); (838.03, 30) (893.14, 20)	(819.56, 50)

(continued)

Table 2. (*continued*)

Instance	Computational results	BKS
	Vehicle: 10; TimeCalc: 39.2 s	Vehicle: 10
VRPNC13	(1784.13, 114); (1786.99, 94); (1788.44, 45); (1791.79, 36); (1793.84, 35); (1798.86, 34); (1802.48, 33); (1829.78, 32); (1834.51, 27); (1837.76, 26); (1889.78, 25); (1909.25, 22); Vehicle: 13; TimeCalc: 67.5 s	(1541.10, 70) Vehicle: 11
VRPNC14	(945.60, 120); (950.24, 110); (960.36, 100); (971.33, 90); (976.32, 80); (982.02, 70); (1005.92, 60); (1056.27, 50); (1062.32, 40); (1062.32, 30); (1065.37, 20) Vehicle: 12; TimeCalc: 44.9 s	(886.37, 140) Vehicle: 11

while the range of vehicle load can be reduced. Also, the total cost of the solutions is relatively similar with the total cost of its corresponding BKS.

VRPNC11-VRPNC14 is the last case study to be solved. This VRPNC has a cluster-like customer structure instead of randomly located, which VRPNC1-VRPNC10 customers have. The proposed framework is also able to generate solution with smaller range of vehicle load, even though additional vehicles are required for VRPNC11, VRPNC13, and VRPNC14.

Solution generated by PSO is basically affected by particles movement, so that movement strategy has an important role to the quality of generated solutions. Another experiment was being conducted to understand how the movement strategy could affect the quality of generated solutions. The experiment was done using VRPNC5 and the PSO parameters values are similar to those of the first experiments. This experiment meant to solve the VRPNC5 instance with several movement strategies. Figure 2 shows the calculation results from the proposed framework of VRPNC5 instance using 6 different types of movement strategy. It is noted that MS0 is the result of original PSO for single objective optimization using similar parameters. While the particles are moved based on the first objective in the MS0, the second objective still can be evaluated for each solution obtained. Therefore, the Pareto front can be generated and updated but not influencing the particles movement in the MS0. It can be concluded from the Pareto front generated from each movement strategy that movement strategy MS1 is able to generate better solution than others whereas the movement strategy MS3 generates the worst solution compared to others.

5   Conclusion

The capacitated vehicle routing problem with load balancing (CVRPLB) is considered in this paper, in which not only minimizing the total cost but also minimizing workload balance among operating vehicles. The multi objective particle swarm optimization (MOPSO) is selected for solving the CVRPLB, by combining and modifying existing implementation of MOPSO [29] and application of PSO for CVRP [18]. The key issues

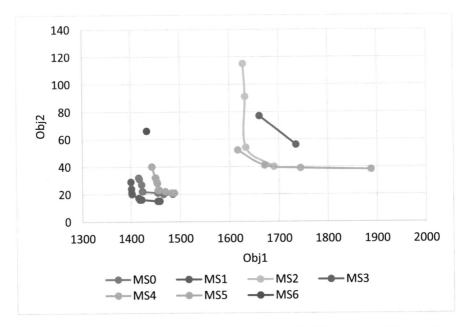

Fig. 2. Pareto front of effect of movement strategies for VRPNC5 solution (Obj1: total cost, Obj2: range of vehicle load)

of the application of MOPSO for CVRPLB are the solution representation, decoding method, and movement strategy of the particles.

Computational experiments using Christofides et al.'s VRPNC data [1] are showing effectiveness of the proposed framework for solving the CVRPLB, in which the framework is able to generate solution that has better load balance than its corresponding best known solution, i.e. providing smaller range of vehicle load than its corresponding BKS. However, this better load balance causes additional in total routing cost and in some situation requires additional vehicles to get the route that serves all customers.

Beside showing that the MOPSO framework is able to solve the CVRPLB, this research also shows that the moving strategy MS1 is the best movement strategy for the MOPSO for CVRPLB case. There are at least two research directions to extend this result, which are exploring the algorithm and extending the application. Detail exploration on the algorithm, i.e. studying the effect of parameter setting, may increase the performance of the MOPSO framework. On the other hand, extending the application to consider more objectives that exist in the VRP domain, i.e. considering number of vehicles as additional objective, may extend the generality of this framework. The decision maker also may have variety of solution alternatives whenever dealing with multi-objective problem situations. Applying the MOPSO framework for tacking another related problem to CVRPLB is also a potential to be conducted.

References

1. Christofides, N., Mingozzi, A., Toth, P., Sandi, C. The vehicle routing problem. In: Combinatorial Optimization. Wiley, New York, p. 431 (1979)
2. Toth, P., Vigo, D. (eds.): The Vehicle Routing Problem. SIAM, Philadelphia (2002)
3. Kumar, S.N., Panneerselvam, R.: A survey on the vehicle routing problem and its variants. Intell. Inf. Manage. 4(3), 66–74 (2012)
4. Laporte, G.: Fifty years of vehicle routing. Transp. Sci. 43(4), 408–416 (2009)
5. Laporte, G., Mercure, H., Nobert, Y.: Exact algorithm for the asymmetrical capacitated vehicle routing problem. Networks 16(1), 33–46 (1986)
6. Lysgaard, J., Letchford, A.N., Eglese, R.W.: A new branch-and-cut algorithm for the capacitated vehicle routing problem. Math. Program. 100(2), 423–445 (2004)
7. Fukasawa, R., Longo, H., Lysgaard, J., De Aragão, M.P., Reis, M., Uchoa, E., Werneck, R. F.: Robust branch-and-cut-and-price for the capacitated vehicle routing problem. Math. Program. 106(3), 491–511 (2006)
8. Clarke, G.U., Wright, J.W.: Scheduling of vehicles from a central depot to a number of delivery points. Oper. Res. 12(4), 568–581 (1964)
9. Gillett, B.E., Miller, L.R.: A heuristic algorithm for the vehicle-dispatch problem. Oper. Res. 22(2), 340–349 (1974)
10. Fisher, M.L., Jaikumar, R.: Generalized assignment heuristic for vehicle routing. Networks 11(2), 109–124 (1981)
11. Berger, J., Barkaoui, M.: A new hybrid genetic algorithm for the capacitated vehicle routing problem. J. Oper. Res. Soc. 54(12), 1254–1262 (2003)
12. Nazif, H., Lee, L.S.: Optimised crossover genetic algorithm for capacitated vehicle routing problem. Appl. Math. Modell. 36(5), 2110–2117 (2012)
13. Augerat, P., Belenguer, J.M., Benavent, E., Corberán, A., Naddef, D.: Separating capacity constraints in the CVRP using tabu search. Eur. J. Oper. Res. 106(2–3), 546–557 (1998)
14. Jin, J., Crainic, T.G., Lokketangen, A.: A parallel multi-neighborhood cooperative tabu search for capacitated vehicle routing problems. Eur. J. Oper. Res. 222(3), 441–451 (2012)
15. Mazzeo, S., Loiseau, I.: An ant colony algorithm for the capacitated vehicle routing. Electron. Notes Dis. Math. 18, 181–186 (2004)
16. Doerner, K.F., Hartl, R.F., Kiechle, G., Lucka, M., Reimann, M.: Parallel Ant Systems for the Capacitated Vehicle Routing Problem. Springer, Heidelberg, vol. 3004, pp. 72–83 (2004)
17. Chen, A.L., Yang, G.K., Wu, Z.M.: Hybrid discrete particle swarm optimization algorithm for capacitated vehicle routing problem. J. Zhejiang Univ. Sci. 7(4), 607–614 (2006)
18. Ai, T.J., Kachitvichyanukul, V.: Particle swarm optimization and two solution representations for solving the capacitated vehicle routing problem. Comput. Ind. Eng. 56(1), 380–387 (2009)
19. Lee, T.R., Ueng, J.H.: A study of vehicle routing problems with load-balancing. Int. J. Phy. Distrib. Logistics Manage. 29(10), 646–657 (1999)
20. Kritikos, M.N., Ioannou, G.: The balanced cargo vehicle routing problem with time windows. Int. J. Prod. Econ. 123(1), 42–51 (2010)
21. Schwarze, S., Voß, S.: Improved load balancing and resource utilization for the skill vehicle routing problem. Optim. Lett. 7(8), 1805–1823 (2013)
22. Hamdi-Dhaoui, K., Labadie, N., Yalaoui, A.: The bi-objective two-dimensional loading vehicle routing problem with partial conflicts. Int. J. Prod. Res. 52(19), 5565–5582 (2014)
23. Carlsson, J.G.: Dividing a territory among several vehicles. INFORMS J. Comput. 24(4), 565–577 (2012)

24. Banos, R., Ortega, J., Gil, C., Marguez, A.L., De Toro, F.: A hybrid meta-heuristic for multi-objective vehicle routing problems with time windows. Comput. Ind. Eng. **65**(2), 286–296 (2013)
25. Banos, R., Ortega, J., Gil, C., Fernandez, A., De Toro, F.: A simulated annealing-based parallel multi-objective approach to vehicle routing problems with time windows. Expert Syst. Appl. **40**(5), 1696–1707 (2013)
26. Lacomme, P., Prins, C., Prodhon, C., Ren, L.: A multi-start split based path relinking (MSSPR) approach for the vehicle routing problem with route balancing. Eng. Appl. Artif. Intell. **38**, 237–251 (2015)
27. Oyola, J., Lokketangen, A.: GRASP-ASP: an algorithm for the CVRP with route balancing. J. Heuristics **20**(4), 361–382 (2014)
28. Martinez-Salazar, I.A., Molina, J., Angel-Bello, F., Gomez, T., Caballero, R.: Solving a bi-objective transportation location routing problem by metaheuristic algorithms. Eur. J. Oper. Res. **234**(1), 25–36 (2014)
29. Nguyen, S., Ai, T.J., Kachitvichyanukul, V.: User's Manual Object Library of Evolutionary Techniques (ET-Lib). Asian Institute of Technology, Thailand (2010)

Some Modifications in Binary Particle Swarm Optimization for Dimensionality Reduction

Shikha Agarwal[(⊠)] and Rajesh Reghunadhan

Department of Computer Science, Central University of Bihar,
Patna, Bihar, India
shikhaagarwal@cub.ac.in, kollamrajeshr@ieee.org

Abstract. Binary particle swarm optimization (BPSO) algorithm is a nature inspired algorithm which has seen many applications in optimization across disciplines. BPSO algorithm consists of mainly equations for velocity updation and position updation. In this paper some of the modifications for binary particle swarm optimization (PSO) on those equations are suggested. The classification results of benchmarking dataset, namely, SRBCT and DLBCL with feature selection using proposed modifications and improved binary PSO (IBPSO) show that they have equal merits.

Keywords: Particle swarm optimization · Classification · Feature selection

1 Introduction

High dimensionality is characterized by very high number of features as compared to the numbers of samples. Need of high dimensional data has become de-facto rule rather than exception in almost all areas of day-today life, for example in medical care, stock market, gene expression microarray data, etc. Therefore high dimensionality brings blessing rather than curse because we are consumers of data. On the other hand for the people who are drawing useful information from junk of data, it is curse rather than blessing due to some challenges [1], namely, three unique characters (noise accumulation [2, 3, 8], spurious correlation [4], incidental homogeneity [4, 5]), empty space phenomena [6], intrinsic dimensionality [7] and concentration of measure phenomenon [8].

Many statistical and computational methods have been proposed by many researches to overcome the problem of high dimensionality. Almost all dimensionality reduction methods when applied with real life ultra high dimensional data, some degradation in the performance occur because of the different challenges that need to be addressed. For effective handling of high dimensional data in both statistical and computational perspective, the use of feature selection has been promoted in many research areas [9]. In a review of Bolon et al. (2014) [10] on feature selection methods concluded with the need of new methods of dimensionality reduction which are able to increase the robustness of selected feature subset of features.

Nature inspired computational methods has been widely used recently for dimensionality reduction. Particle swarm optimization (PSO) [13] developed by Kennedy and Eberhart (1995) is one of the efficient nature inspired algorithm which has seen many

© Springer Nature Switzerland AG 2019
B. K. Ane et al. (Eds.): WSC 2014, AISC 864, pp. 205–212, 2019.
https://doi.org/10.1007/978-3-030-00612-9_18

applications since 1995. Later they developed binary PSO (BPSO) for discrete binary variables [14]. Since these are iterative algorithms, reducing the number of steps will lead to reduction in the computational complexity. Recently Chuang et al. proposed improved binary PSO (IBPSO) [15] with modified velocity and position updating equations.

In this paper four modifications of IBPSO are proposed to reduce dimensionality with K-nearest neighbour (K-NN) [11, 12] method as an evaluator for gene expression data classification problems and the results are compared. This paper is organized as follows. Section 2 deals with a review of improved binary particle swarm optimization. Section 3 presents the four modified IBPSO algorithms. Section 4 deals with the results and discussions. Section 5 concludes the paper.

2   Improved Binary Particle Swarm Optimization (IBPSO) - A Review

Particle swarm optimization (PSO) was developed by James Kennedy and Russell Eberhart in 1995 [13]. PSO is based upon the swarming behaviour of animals and birds. PSO was designed for optimization problem of real value space, but in real life many optimization problems occur in discrete featuring space. Hence, Kennedy and Eberhart (1997) [14] developed the Binary PSO for the discrete binary variables. In binary PSO each particle is represented by binary bits, a vector of 0 and 1. 0, represent the absence of a feature and 1 represents the presence of that feature. Each particle adjust its position according to personal (pbest) and global (gbest) best value. If global best itself is trapped into the local optimima, then search will also end with the local optimum solution.

To avoid being trapped in local optimum, improved binary PSO was developed. IBPSO has been proposed by Chuang in 2008 [15] which uses the concept of resetting the gbest. In IBPSO gbest gets retires under some circumstances like if value of gbest does not changes for continuous three iterations, gbest gets reset to zero. In IBPSO the position of each particle is initialized by randomly generating a binary string of length same as the number of attributes.

The fitness of each particle is calculated based on classification accuracy of 1-NN classifier using leave one out cross validation method. The best accuracy of each particle is stored as personal best fitness and position string corresponding to that fitness is stored as personal best position (pbest). The best of all the pbest fitness and position has stored as global best fitness and position (gbest). Each particle's position and velocity is updated according to the following three equations.

$$v_{pd}^{new} = w v_{pd}^{old} + c_1 rand_1 \left(pbest_{pd} - x_{pd}^{old} \right) + c_2 rand_2 (g_{pd} - x_{pd}^{old}) \tag{1}$$

$$S\left(v_{pd}^{new}\right) = 1/(1 + e^{-v_{pd}^{new}}) \tag{2}$$

$$if \left(rand < S\left(v_{pd}^{new}\right) \right) \; then \; x_{pd}^{new} = 0 \tag{3}$$

where v_{pd} is the velocity of the particle 'p', 'd' indicates the dimension, 'w' is the inertia weight, c_1 & c_2 are acceleration (learning rates) factors and, $rand$, $rand_1$ & $rand_2$ are random numbers. $x_{3,4}^{old} = 0$ means that the 4th position of the 3rd particle is not selected.

After Eq. (1), if the calculated new velocity does not appear in the range of $[V_{min}, V_{max}]$ then the following max_min operation is performed to make it in the limit.

$$if\ v_{pd}^{new} \notin (V_{min}, V_{max})\ then$$

$$v_{pd}^{new} = max(min(V_{max}, v_{pd}^{new}),\ V_{min}) \qquad (4)$$

3 Some Modification in IBPSO

3.1 Modified IBPSO-1

In IBPSO, Eqs. (1) and (3) uses three random numbers. $rand_1$ contribute to the personal model of the particle and $rand_2$ contribute to the social model of the particle. Third random number is used for the comparison with the new transformed velocity.

In place of generating three numbers, one random number can fulfil the need in all the three places. Therefore, replacement of all the three random numbers with a single random number has been made in modified IBPSO-1. Hence the following equation corresponds to the modified IBPSO-1.

$$r_1 = rand \qquad (5)$$

$$v_{pd}^{new} = wv_{pd}^{old} + c_1 r_1 \left(pbest_{pd} - x_{pd}^{old} \right) + c_2(1 - r_1) \left(gbest_d - x_{pd}^{old} \right) \qquad (6)$$

$$S\left(v_{pd}^{new} \right) = 1/(1 + e^{-v_{pd}^{new}}) \qquad (7)$$

$$if\ \left(r1 < S\left(v_{pd}^{new} \right) \right)\ then\ x_{pd}^{new} = 1\ else\ x_{pd}^{new} = 0 \qquad (8)$$

3.2 Modified IBPSO-2

Equations (1) and (3) are based upon the $rand$ function of matlab which generates the numbers between [0,1]. This forces the particle to move in the forward direction only. Matlab also consist of one more function named; $randn$ which generates the number in the range $[-1,1]$. Use of $randn$ in the place of $rand$ function will allow the particle to move in the backward direction also. Modification of Eq. (1) by replacing the $rand$ with $randn$ has been performed in modified IBPSO-2. Hence the following equation corresponds to the modified IBPSO-2.

$$v_{pd}^{new} = wv_{pd}^{old} + c_1 randn_1 \left(pbest_{pd} - x_{pd}^{old} \right) + c_2 randn_2 (gbest_d - x_{pd}^{old}) \qquad (8)$$

$$S\left(v_{pd}^{new} \right) = 1/(1 + e^{-v_{pd}^{new}}) \qquad (9)$$

$$if \left(rand < S\left(v_{pd}^{new} \right) \right) then \ x_{pd}^{new} = 0 \qquad (10)$$

3.3   Modified IBPSO -3

Equation (2) is used to transform the velocity value using the sigmoidal function.

$$S\left(v_{pd}^{new} \right) = 1/(1 + e^{-v_{pd}^{new}}) \qquad (11)$$

If $S(v)$ is less than 0.5, then it indicates about the negative velocity value. In Eq. (3), this transformed value has been compared with the random number *rand*. If $S(v)$ value is greater than random number, then the position feature is accepted (bit is set to 1) otherwise the position feature is rejected (bit is set to 0). Equation (3) is for the selection and rejection of the velocities. Some velocities has positive value and some has negative values. Thus, it means that, $S(v)$ indicates the probability distribution of velocities and *rand* is used to reject the velocities with some probability.

If we compare $S(v)$ directly with 0.5 then it forces the very poor velocity update to get rejected and accept the good update in velocity. Therefore replacement of rand in Eq. (3) with 0.5 has been made in modified IBPSO-3. Hence the following equation corresponds to the modified IBPSO-3.

$$v_{pd}^{new} = wv_{pd}^{old} + c_1 randn_1 \left(pbest_{pd} - x_{pd}^{old} \right) + c_2 randn_2 (gbest_d - x_{pd}^{old}) \qquad (12)$$

$$S\left(v_{pd}^{new} \right) = 1/(1 + e^{-v_{pd}^{new}}) \qquad (13)$$

$$if \left(S\left(v_{pd}^{new} \right) > 0.5 \right) then \ x_{pd}^{new} = 1 \ else \ x_{pd}^{new} = 0 \qquad (14)$$

3.4   Modified IBPSO-4

Use of sigmoidal function shows that, if v has negative value then transformed valuewill also be negative. Therefore in modified IBPSO-4, direct comparison has beenconducted through checking, whether v has negative or positive value. There is no needof transformation of velocity using sigmoidal function. Hence the following equationcorresponds to the modified IBPSO-4.

$$v_{pd}^{new} = wv_{pd}^{old} + c_1 randn_1 \left(pbest_{pd} - x_{pd}^{old} \right) + c_2 randn_2 (gbest_d - x_{pd}^{old}) \qquad (16)$$

$$if \left(v_{pd}^{new} > 0 \right) \ then \ x_{pd}^{new} = 1 \ else \ x_{pd}^{new} = 0 \qquad (17)$$

4 Results and Discussions

Dimension reduction of gene expression data is the common challenge in the field of computational cancer research in biomedicine and molecular biology. A correct analysis using computational power could solve the complex problem of cancer research. Due to the time consuming study on gene expression data, pretreatment of gene expression data using feature selection could be used to reduce the analysis time without increasing the error rate of analysis. Many researchers are working on the two goals namely; dimensionality reduction and decrease in error rate or increase in accuracy.

In this study high dimensional gene expression data has been used for dimensionality reduction to find the minimum number of genes which are able to classify the dataset with maximum accuracy. The parameter setting of the experiment are shown in Table 1.

Table 1. Parameter setting for all the five algorithms

Parameter	Value
Interia weight	0.5
Learning rates	$c_1 = 2$; $c_2 = 2$
Random number	[0–1]
Velocity range	[$Vmin$, $Vmax$] = [−6,6]
Number of particles	40
Number of maximum iteration	150
Stopping criteria	$max \ iteration = 150$ or $max \ accuracy = 100$

The datasets used for the study consist of two gene expression profiles (SRBCT and DLBCL) from http://www.gemssystem.org. The SRBCT data includes neuroblastoma (NB), rhabdomyosarcoma(RMS), non-Hodgkin lymphoma (NHL), and the Ewing family of tumors (EWS). The number of samples are eighty three, number of different categories of tumors are four and number of total genes are 2308.

DLBCL is the gene expression profiles of diffuse large B-cell lymphoma (DLBCL). The number of samples are seventy seven with two different categories of tumors and the total number of genes are 5469.

In this work IBPSO and all the four modifications in IBPSO have been applied ten times on each data set (and averaged) since all the five algorithms consist of random numbers therefore merely relying on the result of one run do not justify the actual dimensionality reduction. The results of all the ten runs of one of the two data sets (SRBCT dataset) have been shown in Table 2. In Table 2 few runs are showing 100% classification accuracy. Run 2 and Run 4 of modified IBPSO3 has 100% classification accuracy. In case of modified IBPSO-4 Run 2, Run 6 and Run 9 shows 100% classification accuracy. This 100% classification accuracy indicates that not all features are needed to acquire perfect classification.

Table 2. Classification accuracy (CA) and the selected number of genes (SG) in ten run of all the four modified IBPSO on RSBCT

RUN	Modification1		Modification2		Modification3		Modification4	
	Accuracy selected genes		Accuracy selected genes		Accuracy selected genes		Accuracy selected genes	
Run 1	98.80	1020	96.39	1152	97.59	229	97.59	440
Run 2	96.39	1035	97.59	1112	**100**	215	**100**	216
Run 3	97.59	1143	96.39	1152	97.59	229	96.39	266
Run 4	96.39	1012	97.59	1112	**100**	215	96.39	397
Run 5	97.59	1030	97.59	1136	96.39	463	97.59	440
Run 6	96.39	1064	96.39	1104	97.59	229	**100**	216
Run 7	98.80	1128	6.39	1152	97.59	229	96.39	266
Run 8	97.59	1134	97.59	1112	98.80	313	97.59	440
Run 9	97.59	1157	97.59	1136	98.80	491	**100**	216
Run 10	97.59	1073	96.39	1152	96.39	133	96.39	266

The result in Table 3 shows the average accuracy and average selected genes in all the five cases. The average results over ten runs in our implementation of the IBPSO [15] gives 97.59% and 94.51% classification accuracy for SRBCT and DLBCL data sets respectively with respective average selected genes as 1123 and 2697. Table 3 shows that modified IBPSO-3 and modified IBPSO-4 have much better dimension reduction power with comparable classification accuracy with respect to IBPSO. Number of selected genes in modified IBPSO-3 and modified IBPSO-4 are 275 and

Table 3. Average classifcation accuracy and selected genes in IBPSO and four modified IBPSO

Algorithm	SRBCT		DLBCL	
	Accuracy selected genes		Accuracy selected genes	
Improved binary PSO	97.59	1124	94.51	2697
Modified IBPSO-1	97.47	1080	95.84	2691
Modified IBPSO-2	96.99	1132	93.64	2734
Modified IBPSO-3	98.05	275	96.10	282
Modified IBPSO-4	97.83	316	96.10	407

316 for SRBCT while for DLBCL selected genes are 282 and 407. For both data sets, number of selected genes using third and fourth modification are much lower than the selected genes by IBPSO.

Table 4 clearly indicates the percentage in dimensionality reduction in both the data sets. Modified IBPSO-3 has showed 88.08% and 94.84% dimensionality reduction in SRBCT and DLBCL datasets respectively.

Table 4. Summary of dimensionality reduction using IBPSO-3 and IBPSO-4

	SRBCT		DLBCL	
	IBPSO-3	IBPSO-4	IBPSO-3	IBPSO-4
Total genes	2308	2308	5469	5469
Average selected genes	275	316	282	407
Average classification accuracy	98.05	97.83	96.10	96.10
Average dimension reduction	88.08%	86.30%	94.84%	92.55%

5 Conclusion

In this study, improved binary particle swarm optimization (IBPSO) is used to implement four modifications in IBPSO for the better dimensionality reduction and classification of gene expression data using K-nearest neighbour (K-NN) evaluator. Experimental results shows that out of four modifications, third and fourth modifications have effectively reduced the total number of genes (features) needed for the satisfactory classification accuracy. The classification accuracy obtained by the proposed methods was almost same as that of IBPSO. The percentage of dimensionality reduction for the proposed third modification for SRBCT and DLBCL are 98.05% and 96.10% respectively. Similarly for the fourth modification, percentage of dimensionality reduction in case of SRBCT and DLBCL are 97.83% and 96.10%. The proposed modifications can serve as pre-processing method so that analysis time could be reduced without increasing the error rate of analysis. It could serve as a tool for dimensionality reduction in translational application across the fields. Further studies are going on with more data sets to obtain minimum number of attributes and satisfactory classification accuracy in translational applications.

References

1. Fan, J., Han, F., Liu, H.: Challenges of big data analysis. Nat. Sci. Rev. **1**(2), 293–314 (2014)
2. Hastie, T., Tibshirani, R., Friedman, J.: The Elements of Statistical Learning. Springer (2009)
3. Buhlmann, P., Geer, S.V.: Statistics for High Dimensional Data: Methods Theoryand Applications. Springer (2011)
4. Fan, J., Liao, Y.: Endogenity in high dimensions. Ann. Stat. **42**(3), 872–917 (2014)

5. Liao, Y., Jiang, W.: Posterior consistency of nonparametric conditional momentrestricted models. Ann. Stat. **74**(1), 37–65 (2011)
6. Scott, D.W., Thompson, J.R.: Probability density estimation in higher dimensions. In: Gentle, J.E. (ed.) Computer Science and Statistics: Proceedings of the Fifteenth Symposium on the Interface, pp. 173–179. Amsterdam, New York. Oxford, North Holland-Elsevier Science Publishers (1983)
7. Fukunaga, K.: Introduction to Statistical Pattern Reconition. Academic Press Professional, San Diego (1990)
8. Donoho, D.: High-Dimensional Data Analysis - The Curse and Blessings of dimensionality. AMS Math Challenges Lecture, pp. 1–32 (2000)
9. Fodor I.: A survey of dimension reduction techniques, U.S. Department of Energy (2002)
10. Bolon-Canedo, V., Sanchez-Maronpo, N., Alonso, A.: A review of microarraydatasets and applied feature selection methods. Inf. Sci. **282**, 111–135 (2014)
11. Cover, T., Hart, P.: Nearest neighbor pattern classification. In: Proceedings of the IEEE Transactions Information Theory, pp. 21–27 (1967)
12. Fix, E., Hodges, J.L.: Discriminatory Analysis-Nonparametric DiscriminationConsistency Properties. Technical Report 21-49-004, Report no. 4 (1951)
13. Kennedy, J., Eberhart, R.C.: Particle swarm optimization. In: Proceedings of the 1995 IEEE International Conference on Neural Networks, vol. 4, pp. 1942–1948 (1995)
14. Kennedy, J., Eberhart, R.C.: A discrete binary version of the particle swarm algorithm. In: Proceedings of the IEEE International Conference on Computational Cybernetics and Simulation, vol. 5, pp. 4104–4108 (1997)
15. Chuang, L.Y., Chang, H.W., Tu, C.J., Yang, C.H.: Improved binary PSO forfeature selection using gene expression data. Comput. Biol. Chem. **32**(1), 29–38 (2008)

Some Thoughts of Soft Computing

Collaborative Project Management Framework for Partner Network Initiation in Machining Domain

Eduard Shevtshenko[1(✉)], Igor Polyantchikov[2], Kashif Mahmood[1],
Taivo Kangilaski[1], Alexander Norta[1], Tatjana Karaulova[1],
and Ardo Perm[3]

[1] Tallinn University of Technology, Tallinn, Estonia
{eduard.sevtsenko,kashif.mahmood,alex.norta,
tatjana.karaulova}@ttu.ee,
taivo.kangilaski@energia.ee
[2] Densel Baltic, Tallinn, Estonia
poljantsikov.igor@gmail.com
[3] Apeco, Saue, Harju, Estonia
ardo@apeco.ee

Abstract. Many manufacturing companies are facing problems related to the initiation and management of collaborative manufacturing projects. Collaboration between small- and medium-sized enterprises (SMEs) enables business partners to share equipment, labour, and skill-sets to fulfil larger projects in Partner Network (PN). We define PN as a group of companies that work together to enhance the capabilities of each other. The start of a new project triggers the establishment of a Virtual Organisation (VO). In this research paper, we present the framework for Collaborative Project Management in VO that includes the following collaboration-process steps: project initiation, project-plan preparation, searching for a subcontractor, preparation of quotations, evaluation of PN proposals, project-proposal formations, PN-agreement elaboration, and project management. In the current paper, the authors suggest using the novel five level modeling approach for Collaborative Business Process (CBP) modeling. Authors are using the Value-Added Chain (VAC) and Event Process Chain (EPC) notations to design the CBP in the ARIS architect software that enables to visualize the inputs, outputs and relationships between high level and low-level business processes. Project management in a VO is more sustainable then in conventional supply chains due to the common tools that enable the measurement of project success in a unified way with the possibility of responding quickly to changed conditions.

Keywords: Partner network · Collaborative business process · ARIS
VAC · EPC · E-sourcing · Decentralized autonomous organizations

1 Introduction

With increased business competition, companies strive to gain advantages by forming networks of enterprises. Such collaborative networking triggers improvements in business processes because of employing new tools and methods, which allow for

© Springer Nature Switzerland AG 2019
B. K. Ane et al. (Eds.): WSC 2014, AISC 864, pp. 215–233, 2019.
https://doi.org/10.1007/978-3-030-00612-9_19

developing competencies, combine strengths of participating companies and discarding weaknesses. The management of collaborative projects in the partner network (PN) requires clear definitions of project goals and business alliances in PNs that contribute to a proper initiation, planning, execution, monitoring and controlling of projects. At the same time, such a streamlined approach helps to achieve project objectives. To overcome the difficulties that manufacturing companies face in starting a VO-collaboration with other companies and work effectively, a collaborative project management process requires detailed definitions of target improvements. Collaborative Project Management (CPM) offers a bifurcation of projects into different smaller parts so that actively participating members in a project understand the process map. The possible sub-parts or sub-plans connect to each other and synchronize the collaboration flow so that members of a PN are aware of the common collaboration purpose.

This paper aims to develop a process model (framework) for collaborative project management in PN that facilitates for SMEs the entry to collaborative networks. The collaboration requirements we express in the form of key performance indicators (KPIs) and mandatory project-process steps. The paper gives a position to collaborative project management with BPM (Business Process Management). For the problem description, the authors follow a CPM framework for collaborative project management that focuses on horizontal collaboration. The latter means the partners have different competences and do not compete with each other in a project.

The structure of the paper is as follows. First, Sect. 2 discusses related work, followed by Sect. 3 where the business-process modelling for PN-organisations is presented. Next, Sect. 4 gives a case study for collaborative project management. Finally, in Sect. 5 we give a conclusion and discuss open issues for future work. Future research will be dedicated to the vertical collaboration, where the partners-competitors are providing similar services or resources.

2 Related Work

For introducing the reader to the necessary concepts and frameworks to understand the remainder of the paper, we discuss in Sect. 2.1 collaborative project management, followed by defining the notion of a project in Sect. 2.2. Next, the project-phases explanations in Sect. 2.3 are followed by specifics for project management in Sect. 2.4. Finally, mathematical methods for CPM in Sect. 2.5 precede a business-process methodology explanation in Sect. 2.6, and a CPM problem description terminates in Sect. 2.7.

2.1 Collaborative Project Management

The discipline of project management increases in importance for organizations that adopt business-processes awareness. Additionally, project management is the essential activity for enterprises. Project management comprises a dynamic set of tools that improve the ability of humans to plan, implement and manage activities for accomplishing specific organizational objectives. On the other hand, project management is not just a set of tools, but it is a result-oriented management approach, that facilitates the formation of collaborative relationships among a various sort of characters [1].

Project management is also a crucial facilitator that enables a business improvement methodology that is similar to six sigma and lean manufacturing that companies adopt to improve their efficiency and competitiveness. In fact, robust project management indicates a company's core competence and allows for exploiting the effects of these improvement methodologies [2].

2.2 Defining a Project

A project is not a day-to-day operation, but a routine or repetitive work; it is neither the size nor the length of a methodology that determines whether it is a project or not, but it is something specific and limited in time. Projects can be further defined as temporary rather than permanent social systems or work systems that are constituted by teams within or across organizations to accomplish particular tasks under time constraints [3]. A project is a work method or methodology that strongly focuses on the goal. The project needs to be time bounded and have proper resources to carry it forward. Moreover, the project goals should be described, and the budget must be declared in project initiation. A project can be defined as:

A unique assignment that has a specific goal, time oriented that means the specific period, own budget that corresponds to specific resources and a unique work arrangement – temporary organization [4].

According to Gray and Larson [1] – a project is a complex, non-routine, one-time effort limited by time, budget, resources, and performance specifications designed to meet customer needs.

The most important parameters for evaluating a project success are quality, time and resources or scope, time and cost. These triple constraints represent a project management triangle and prioritize according to their significance with one being predominant, e.g., whether it is the quality of a product, the cost of resources, or delivery time that is essential to a project. The product quality indicates how well the project will be served up to the certain desired level, while the calendar time corresponds to the project duration. Finally, resources are money, numbers, or labor hours [4].

2.3 Phases of Project

Projects comprise phases that are organized according to a project lifecycle of four phases; pre-study (defining), planning execution and closure. Figure 1 depicts such a general project model. To ensure that a project is followed up and evaluated the model can also include impact realization and verification of a business impact goal using risk analysis. The project model in Fig. one is simple and easily compatible with specific project management models without any difficulty. Other companies and enterprises follow distinct project management models, and this general model facilitates communication among individuals involved in particular for project realisation. Moreover, there are other common project management models, but the primary difference is the number of stages and terms are changed. Otherwise, the structure and contents pursue the general project model in principle. Each phase of the project is a further refined process and a small sub-project so that each phase must be initiated, planned, executed, closed, managed and evaluated.

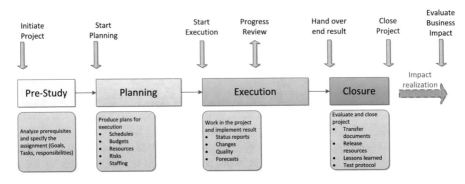

Fig. 1. Generic project model [3]

2.4   Project Management in Partner Networks

Increased market competition force SMEs to collaborate with other enterprises to compete successfully against larger corporations. Such networks are known from literature as Industrial Clusters, Partner Networks (PN), Collaboration Networks (CN), Collaborative Networked Organizations (CNO), Virtual Breeding Environment (VBE) etc. [5]. A partner network (PN) is a collaborative network of enterprises, where partners agree on collaboration rules in signed agreements. The collaboration between PN enterprises is organized by collaborative projects or virtual projects, where all partners share common goals: an increase of profitability, increase of the revenue, optimal usage of resources, on-time delivery, lowest prices to the customers, increase of the customer database, fast growing turnaround, providing better quality, and the decrease of total lead-time [6]. To be able to achieve these goals, authors apply the project-management (PM) methodology for a cross organizational projects, which are known as Collaborative Project Management (CPM) and received attention by researchers. Aekyung et al. [7] have proposed the CPM methodology for monitoring and analyzing the collaborative performance of partners based on certain key performance indicators (KPI). Kwangyeol et al. [8] introduce i-Manufacturing and collaboration systems with functions and goals, see Table 1.

Ruy et al. [9] suggest the collaborative process modelling (CPM) method, transform it into Petri Nets models and develop the fractal manufacturing system (FrMS) architecture [10]. In collaborative project management architecture, Chen et al. [11] emphasize that the close monitoring of the PM process enables efficient and effective change management. Additionally, project measures are important for the assessment of the wellbeing of a PM process.

Table 1. The goals of i-Manufacturing project [8]

Goals	Objectives	Plans
Informatization	By merging IT, knowledge and communication technologies, to achieve innovation and integrate all manufacturing entities to build up a common infrastructure supported by the facilitation of inter- and intra technological collaboration.	• Build up collaboration models and networks to support technological collaboration within companies. • Promote to SMEs the use of collaboration systems by informing about successful cases. • Enlarge application areas to other industrial sections and regions. • Integrate the infrastructure including legacy systems, etc.
Innovation in manufacturing processes	Through the integrative management of manufacturing information and the interoperability of manufacturing processes between companies, to improve the ability to cope with market demand, with optimized and improved manufacturing processes.	• Develop technologies for improving interoperability between various kinds of processes focusing on collaboration. • Optimize manufacturing processes. • Synchronize the manufacturing process (MP) with business process (BP). • Add enhanced knowledge into processes by applying new technologies such as IT/UT.
Innovation in manufacturing systems	To enable companies to cope with mass-customization by developing a distributed and reconfigurable manufacturing system and synchronizing the system with manufacturing information.	• Develop a highly intelligent and distributed manufacturing system enabling self-reconfiguration. • Investigate operational technologies for synchronizing components in the manufacturing system. • Build up and operate a pilot plant to verify system performance.
Innovation in developing new products	To improve product quality and reduce delivery time of brand-new products, thereby enhancing competition power in global markets by building up a synthetic service center especially for SMEs.	• Build up a synthetic support center for supporting the development of brand-new products by SMEs. • Develop effective and efficient strategies for operating centers.

2.5 Mathematical Methods for CPM

There are many mathematical methods and techniques for the selection of suppliers from alternatives that are based on qualitative or quantitative factors. The extensive study of the problem shows several directions for supplier selections in multiple criteria decision making. Techniques are used in different mathematical methods, such as the analytic hierarchy process (AHP), analytic network process (ANP), case-based reasoning (CBR), data envelopment analysis (DEA), fuzzy set theory, genetic algorithm (GA), mathematical programming, the simple multi-attribute rating technique (SMART), Lagrangian relaxation and their hybrids. [12] These techniques are classified by two main directions, first a consideration of quantitative perspectives and second, the use of qualitative perspectives.

Quantitative techniques evaluate such factors as price, time, distance, profit and use trivial arithmetic calculation. They lead to the appropriate result seen from a financial perspective, but not necessary the best solution from a manufacturing or a customer point of view [13].

The second direction of partner selection methods deal with a qualitative perspective and considers quality, skill, flexibility and competence as criteria. In practice, the qualitative decision-making is based on the subject impressions of decision makers. Often, the decision making is not fact-based but employs an intuitive human judgment and relations between companies. In some cases, the decision is difficult and clarified by the decision maker himself. Qualitative partner evaluation deals with multi-criteria, uncertainty and data fuzziness that enforce the use of mathematical tools that must take into account these facts. The theory of Fuzzy Set is applied in this case. There is a variety of related frameworks that deal with fuzzy theory issues that support cases where uncertainty is involved.

Some methods aid the decision-maker to consider both qualitative and quantitate factors in the calculation. "The methodology uses fuzzy QFD (quality function deployment) to convert qualitative information into quantitative parameters and then combines this data with other quantitative data to parameterize a multi-objective mathematical programming model." [14] There exist methods that consider only a specific problem or limited problem in a particular area. Talluri [15] developed a method for selecting a partner in the formation of a VO that uses a two-phase mathematical programming approach for efficiency partner selection. O'Brain [16] applies an integrated approach for AHP and linear programming that considers both qualitative and quantitative factors in selecting the best suppliers. Hwang and Yoon [17] present TOPSIS (a technique for order preference by similarity to an ideal solution), a multi-criteria method to propose a problem decision from some alternatives. The principle is considering both a positive-ideal and negative-ideal solution in a multi-attribute decision-making problem in that the chosen alternative should have the shortest distance from the positive ideal solution and the farthest distance from the negative ideal solution [18].

Other authors combine the existing supplier selection methodologies and adopt them for PN needs. For this purpose, the authors suggest developing the Partner Efficiency Index (PEI) to consider qualitative and quantitative perspectives to make the selection process more feasible and applicable for focal players.

2.6 Business Process Management (BPM) Methodology

Business process management (BPM) has been broadly accepted in today's organizations. BPM supports businesses by providing a set of tools, methods, and techniques to identify and discover business processes, also to analyze these processes to find opportunities for improvement, to implement the improved processes, and to monitor and control their execution. A business process typically involves different organizational aspects, ranging from human resources to business documents and technology [19]. According to Chinn [20], Business Process Reengineering (BPR) emphasized a holistic focus on business objectives and how processes related to them, encouraging a full-scale recreation of processes rather than an iterative optimization of sub-processes. Jakoubi [21] discusses that one of the important factors in process modelling is risk identification. The integration of BPM and risk management facilitates an organization to sustain and accomplish secure business processes. It can also enhance the ability to reduce risk in business processes by design and to mitigate such risks at the run time.

2.7   CPM Problem Description

Today's SME enterprises are working in an old fashion way: business processes are not documented by managers, IS tools are rarely used for collaboration establishments, the project management activities are rarely supported by analytical tools that enable efficient measurements. The initiation of new collaborative projects and the project-proposal preparation takes the considerable time that leads to a loss of business opportunities.

Collaborative Project Management (CPM) is a complicated task, and the aim of the ongoing research is to suggest the CPM framework for the sustainable realization of collaborative projects. The phases of joining the collaborative network, business processes description and project realization are covered by a suggested conceptual model. The aim of this research is to develop a CPM model for the sustainable realization of collaborative projects for SMEs, apply it in PNs, and describe the second step of the model, CPM business process, in a BPM application software.

3   Conceptual Model of Sustainable Collaborative Projects for SMEs

In a current research paper, authors cover the second step of a Conceptual model for a sustainable realization of collaborative projects for SMEs ion the machinery domain see Fig. 2.

The conceptual model consists of four steps:

- Step 1: Preparation of candidate enterprises to join a PN. To be able to join a PN, the candidate enterprise must fulfill four sub-steps:

  a. Fill in the questionnaire. Based on the questionnaire, it is decided in which area enterprises provide services (related to the business processes of an enterprise), and the enterprise data is added to an appropriate category [23].
  b. Passing a PN audit, that measures organizational maturity [24]. Organizational maturity means that an enterprise describes their business processes based on quality standard (e.g. ISO, EMAS) requirements as BPMN or EPC models [25].
  c. Signing of contractual and Service Level Agreements (SLA).
  d. Filling in the forms for enterprise data transformation towards PN database tables about machine centers and their availability.

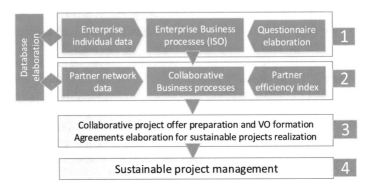

Fig. 2. Conceptual model for the sustainable realization of collaborative projects [22]

- Step 2: The Partner Efficiency Index (PEI) enables the selection of the optimal partners based on a fuzzy preference relation matrix [26] and support the measurement of collaborative business-process efficiency. The PN establishment and the creation of PEI process consist of two sub-steps:

  a. To assess the enterprise readiness to join a PN, the maturity of a collaborative BP and processes of data collection for PN are audited by independent experts after the enterprise business processes are mapped to the collaborative business processes of PN. The input to collaborative business processes is updated automatically.

  b. To rate the partner's enterprises, the PEI is used. It is calculated by a fuzzy tool for supplier evaluation that is adopted for the PN-enterprise selection for VO formation.

- Step 3: The Initiation of collaborative projects consist of collaborative project offer preparation, VO formation and an agreements elaboration process.

The collaborative project (CP) steps to address the preparation and partner selection for VO formation. After receiving the new order request from one of the PN enterprises, the collaborative project initiator, or focal player (FP) selects the required domain, defines the project steps and describes the production route (as in the case of Manufacturing domain). Next, describing the FP precedes selecting the routing operation for suitable partner selection based on resources specifications and resource availability data. The Fuzzy processing mechanism of CPM model calculates Partner Efficiency Index (PEI) for each operation of production route, presented by a candidate partner. Subsequently follows a similarity analysis to manufacture the product sub-assemblies in one location. The output of a similarity analysis is the rating list of possible enterprise partners based on FP preferences for a particular project. The AHP methodology is used for a pairwise comparison of the project price and time importance before the final calculation of the PEI for possible partner enterprises.

Finally follows an a agreements elaboration for sustainable project realization. After all stages of setting up a collaborative project, the candidate partners receive offers from an FP. The FP sorts the candidate list based on a combination of previous experiences of collaboration within a PN and a PEI. After confirming the candidate partner's proposals by an FP, the VO formation takes place. The FP calculation uses collaborative project measures and finally sends the results as a proposal to the customer. After accepting the proposal by a project customer, the customer and FP sign the contract. After that, the FP and the selected VO partners sign the collaboration project contracts.

Step 4. Sustainable collaborative project management. After the CP starts, it is important to provide the environment for successful implementation. This environment enables the tracking of project progress, based on input data that VO members submit.

4 Business Process Modelling for PN Organisations

Conventional process modelling approaches focus on the sequence of processes [26, 27], but the current research project is innovative and directed to organizations that operate in a dynamic environment. Organisations that participate in PN as well as in VO have their own goals, indicators, processes, organizational culture and maturity level (e.g. estimated based on Capability Maturity Model Integration, Lean Six Sigma). Thus, the authors have selected EPC/VAC notations (see Table 2) based on the 4 + 1 level approach for CPM business process modelling:

1 - Enterprise Process Map Value-Added Chain (VAC) model
2 - Process level VAC model includes Sub-procedures
3 - Sub-process level Event Process Chain (EPC) model includes Activities and Responsible
4 - Activity level EPC model includes operations
5 - Level or IT Application views related to IS integration.

To motivate the SMEs from the machinery domain to describe the existing business processes, the authors suggest working out the best practice business processes templates for the machinery domain. The purpose is to simplify and speed up the ISO 9001 certification for new comers and support the dynamic improvement of production quality systems and the current business processes maturity for already certified SME partners. The maturity of internal business processes will also enable the enterprise to be more efficient in a partner network.

From a process modelling perspective, ARIS is a tool that supports several processes modelling notations including BPMN, EPC/VAC etc. Different companies have different approaches in ARIS (e.g. LEGO System A/S, Fortum Heat and Power, Statoil, Vestas Wind Systems A/S), but there was no rule-set that has been previously developed for building up models in ARIS. The authors propose to use the developed modelling approach, which enables to document the organizational structure considering the corresponding roles and processes with inputs and outputs that are associated with certain documents if they exist. For this purpose, the research team has performed test, to determine the most readable notation for business users. From process modelling perspective, ARIS is a tool that supports several processes modelling notations including BPMN, EPC/VAC etc. As a result, the IDEF0 notation, initially designed for the modelling of manufacturing processes is too complex for our purpose. There was

Table 2. VAC/EPC objects used for business process modelling in PN organisation

Symbol	Name / Description
Supplied Goods	Product/Service – illustrate process inputs and outputs
∧ ✕ ∨	Moreover, OR and eXclusive OR (XOR) predicates
Sales Manager	Role or Person type – describes the responsible role
Marketing	Value Added Chain – name of sub processes
Partner inquiry	Document – document used in the process
KPI instance	KPI (Key Performance Indicator) – KPIs related to the process or sub-process
Material sort defined	Event – occurs before and after process steps and activities
Similarity analysis engine	Application system – relation to system or application

consideration between BPMN and EPC/VAC. From a business, perspective the BPMN as well as UML is too complex for business users and considered too IT related to model all aspects needed to describe. Also, the BPMN proposes swim lane type models, while customers prefer an approach that does not swim lane based [28, 29].

5   Collaborative Project Management Case Study

To validate the suggested conceptual model authors provide a case study from the machining domain. In the current case study the collaborative network, include three SMEs partners: FC player- a machinery company that provides the transportation device development and production services to the end consumer; a design company that provides the design services and technical documentation preparation services for the PN and manufacturing company. The latter provides the machining services for the PN.

We suggest using the ARIS modelling environment for collaborative project management in the conveyor product that, which follows a classical business model.

5.1   First-Level CPM Business Process VAC Diagram

The collaborative business process starts from *customer enquiry* that is the input to the *Marketing* process. The latter process results from *Order data*, which is input for the *Sales* business process. The *Order specification* is the output of the *Sales* business process and is an input to *Design* business *process*. It generates the service/material specifications that are consumed by the *Purchase* business process and results in *Technical assignments* that are used by the *Manufacturing* business process together with the *Technical documentations and semi-finished goods* delivered by the *Purchase* process. The *Manufacturing* business process generates *Financial and Shipment doc-uments* consumed by the *Logistics* process. The output of the *Logistics* business pro-cess is *Supplied goods* as shown in Fig. 3. After the FP Company receives an inquiry from the customer, the project manager proceeds accordingly to the PN business process.

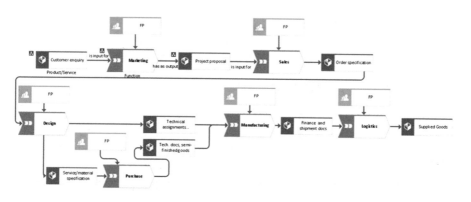

Fig. 3. VAC model of partner network business processes

5.2   Second-Level CPM Business Process Marketing VAC Diagram

FP initiates the project planning process. During the Project Proposal Preparation, the FP decides to outsource the Design and Manufacturing of metal frame parts of the conveyor.

The second level VAC diagram documents the sub-processes with their inputs/outputs and the sub-process owner's managers (Fig. 4). The *Marketing Process* starts with the *Review of customer requirements* Business sub-process. The customer starts this sub-process by sending a quotation to PN. The *Review of customer requirements* sub-process leads to *Material supply*, *Internal resources planning*, *Partner selection* and Logistics planning sub-processes managed by the *Project Managers* of FP or a partner organization. The final step of the *Marketing process* is the *Project Proposal preparation* sub-process that results in the submission of *Project proposal* to the customer.

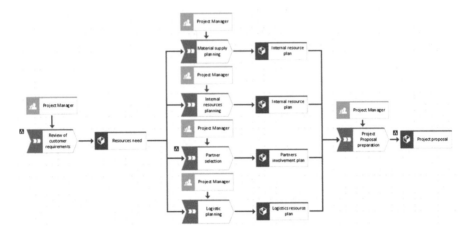

Fig. 4. Marketing sub-process VAC business model

5.3   Third-Level Review of Customer Requirements EPC Diagram

When FP starts the *Review of customer requirements* process, it is described at the Third level by an EPC diagram. The third modelling level is used to document sub-process steps and to determine their sequence. It shows where the process value is generated and where the external inputs are used. From an auditing perspective, it is convenient to show process steps where auditing actions should take place (tollgates). Each sub-process step is assigned to a certain role that identifies the managerial responsibility. That level is modelled by Event Process Chain (EPC) diagram models. The Analysis of the quotation EPC diagram is introduced in Fig. 5. The *Review of customer requirements* sub-process is initiated by the set of documents received together with customer quotation. The *Sales*

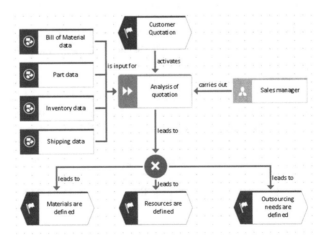

Fig. 5. Project planning sub-process EPC business model

Manager carries out the *Analysis of quotation* sub-process that leads to the *definition of Materials, Resources and Outsourcing needs.*

The Partner Selection EPC diagram is introduced in Fig. 6. The project is started from the *Outsource tasks definition,* and the *Sales Manager* defines the particular project preferences for the partner selection purpose, followed by a similarity analysis and partner inquiries preparation. After the *Sales Manager* has received and verified the quotations, he starts the *Fuzzy processing* to support the partner's selection sub-process. After that, the *Sales Manager* performs the *Risks evaluation* sub-process and *Partner validation* that results in signed agreements. The risk evaluation process is supported FP activities with the purpose of minimizing project risks.

5.4   Fourth-Level Fuzzy Processing Sub-process EPC Diagram

The operations that a decision maker of FP needs to perform are given at the forth level EPC diagram. The Fuzzy Processing sub-process diagram includes operations need for partner assessment. In a current case study, the FP calculates the PEI index to assess the selected partners for outsourcing of a conveyor design and manufacturing operation of the metal frames.

The fourth modelling level covers activities that include the additional operations. This level is designed to show how certain sub-process steps are performed on a further detailed level, where the inputs/outputs are documents/products/services, related business rules, process participants, resources and IT applications.

Thus, this layer is quite an information intensive. Frequently, if processes are not simple, this modelling layer ramains out of scope. However, if the modeler needs to add job descriptions for positions, then it is mandatory to model this level. The *Focal Player* initiates the *Fuzzy Processing* sub-process and the formation of the *Decision-making* committee *formation.* After that, the committee performs *Identify evaluation criteria* for particular project partners. *Decision maker Choose Importance weight* of each criterion and the *Linguistic ratio for Supplier* are activated and followed by the *Calcultion System Aggregate the importance weight of criteria's.* The fuzzy logic based calculation system calculates the PEI used by decision makers for a partner assessment. Based on PEI calculation, the system defines the *distance between each partner* and also the *closeness coefficient.* Based on this information the FP ranks the partners as shown in Fig. 7.

5.5   Virtual Organisation Formation IT Application Principle

A Collaborative Project (CP) offers preparation and partner's selection mechanisms for VO formation are given in Fig. 8. After receiving the new order request from a PN enterprise, the collaborative project initiator, or focal player (FP) selects the required domain, defines the project steps and describes the operation to be fulfilled. In our case study, the FP describes the operations of *Design and Manufacturing domains* that are required for a conveyor. After the FP described the routing operation, suitable partners are selected based on the resources specifications and resource availability data. Candidate partners present the PEI calculation for each operation of a production route. The latter is followed by a similarity analysis to manufacture the product subassemblies

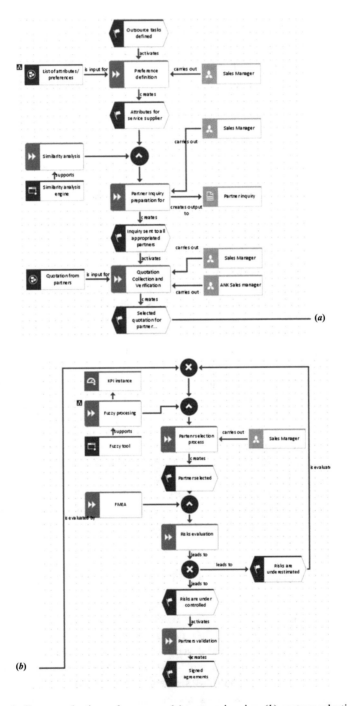

Fig. 6. Partner selection sub-process: (*a*) partner inquiry, (*b*) partner selection.

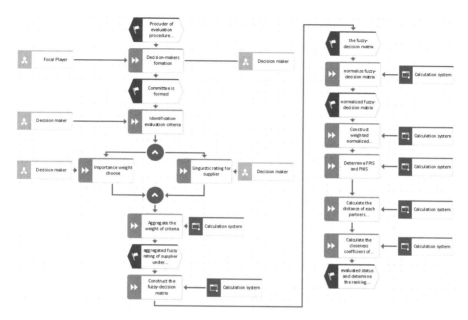

Fig. 7. Fuzzy processing sub-process

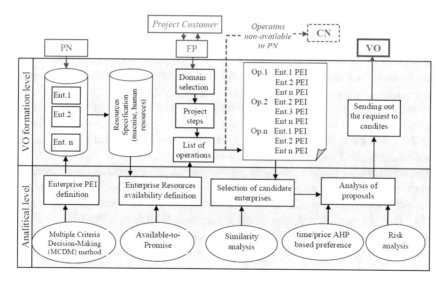

Fig. 8. IT model of virtual organisation formation application

in one location. The output of a similarity analysis is the list of possible enterprise partners that rate best based on the FP preference for a particular project. The AHP methodology allows for a pairwise comparison of the project price and time importance before the final calculation of the PEI for possible partner enterprises.

5.6   Agreements Elaboration for Sustainable Project Realization

After completing all stages of setting up a described collaborative project, the candidate partners selected for design and manufacturing operations receive the offers from an FP. The latter modifies the list based on a combination of the previous experience of collaboration within a PN and PEI. After an FP confirms the VO-candidate partner's proposals, the partners form the VO. The calculation system supports the process of partner selection by FP based on collaborative project measures and the FP finally send a proposal to the customer. After a customer accepts the proposal, the customer and FP sign the contract. Additionally, an FP and the selected VO partner respectively also sign the contract. Those contracts guaranty that the service is paid to all partners after the project is a completion.

6   Conclusion

The result of the current research is a development of a conceptual model for the sustainable realization of collaborative projects. The PN partner selection and proposal preparation activities are covered by the second step of the model. The developed model gives a strategic interpretation of the PN and provides a direction for the development of algorithms for partner efficiency measurements. The current research is limited due to the following reasons: a prototype does not exist; the current level of details is not enough for prototype building. To validate the conceptual model, the approach is applied in a case study from the classical business model.

Business process awareness- makes business collaboration more dynamic, enables to be present in an e-market, facilitates the adoption of service-cloud computing for enterprise needs and decreases the unneeded direct involvement of middle management personal into a collaboration process establishment. At the same time, this research decreased the number of paper work enterprises must conduct to proceed with the composition of a collaborative project proposal. The approach is very powerful for enterprises; however, the authors see the resistance to change from humans as a negative consequence of current research implementation.

The collaborative project management framework is the essential tool that increases the competitiveness of a company starting from the PN initiation, until the tracking of collaborative projects. The calculation system supports the FP decision-making process in CN. This framework enables SME enterprises to act as FP. Additionally, the enterprise that has the order to use the PN resources for fast preparation of customer proposal and to the establishment of collaboration after the customer has accepted the proposal.

The suggested concept is verified under the framework of "collaboration enhancing sustainable conceptual model development and implementation for the SMEs in the machinery domain" that enabled enterprise managers, research staff and PhD. Students to work as one team.

If enterprises accept the idea of this research paper then in future research work, the authors will introduce how to compete successfully with large lean and agile companies.

Acknowledgements. This research was supported by Estonian Ministry of Education and Research for targeted financing scheme B41.

List of Abbreviations:

AHP	- Analytic Hierarchy Planning
BPMN	- Business Process Management Notation
BPM	- Business process management
CNO	- Collaborative Networked Organizations
CP	- Collaborative Project
CPM	- Collaborative Project Management
EPC	- Event Process Chain
FP	- Focal Player
FrMS	- Fractal Management System
IDEF0	- Integrated Definition; method 0
KPI	- Key Performance Indicator
PEI	- Partner Efficiency Index
PN	- Partner Network
VAC	- Value-Added Chain
VBE	- Virtual Breeding Environment
VO	- Virtual Organization
UML	- Unified Modeling Language

References

1. Gray, C.F., Larson, E.W.: Project Management – The Managerial Process, 4th edn. McGraw-Hill, New York (2008)
2. Eve, A.: Development of Project Management Systems, Industrial and Commercial Training, vol. 39, No. 2, pp. 85–90 (2007). Emerald Group Publishing Limited
3. https://en.wikipedia.org/wiki/Project#cite_note-1. Accessed 18 Jan 2015
4. Tonnquist, B.: Project management: a guide to the theory and practice of project. In: Program and Portfolio Management and Business Change. Bonnierutbildning AB, Stockholm (2008)
5. Kangilaski, T., Shevtshenko, E.: Dynamics of partner network. In: The 23rd IEEE International Symposium on Industrial Electronics (ISIE), Bogazici University, Department of Electrical and Electronics Engineering, 34342 Bebek Istanbul/Turk. Istanbul, Turkey, pp. 105–110, 1–4 June 2014
6. Karaulova, T., Shevtshenko, E., Polyanchikov, I., Sahno, J.: Reorganisation of production system on SME enterprises. Ann. DAAAM Proce. **20**, 869–870 (2009)
7. Kim, A., Jung, K.K.J.Y., Shin, D., Kim, B.: Business process warehouse for manufacturing collaboration. In: Proceedings of the 41st International Conference on Computers & Industrial Engineering (2011)
8. Ryu, K., Shin, J., Lee, S., Choi, H.: i-Manufacturing project for collaboration-based Korean manufacturing innovation. In: Proceeding of PICMET 2008, Cape Town, South Africa, 27–31 July 2008

9. Ryu, K., Yücesan, E.: CPM: A collaborative process modeling for cooperative manufacturers. Adv. Eng. Inform. **21**, 231–239 (2007)

10. Ryu, K., Son, Y., Jung, M.: Modeling and specifications of dynamic agents in fractal manufacturing systems. Comput. Ind. **52**, 161–182 (2003)

11. Chen, F., Nunamaker, J.F., Romano, N.C., Briggs, R.O.: A Collaborative Project Management Architecture (2003)

12. Ho, W., Xu, X., Dey, P.K.: Multi-criteria decision making approaches for supplier evaluation and selection: a literature review. Eur. J. Oper. Res. **202**(1), 16–24 (2010)

13. Hvolby, H.-H., Steger-Jensen, K.: Technical and industrial issues of Advanced Planning and Scheduling (APS) systems. Comput. Ind. **61**(9), 845–851 (2010)

14. Erol, I., Ferrell, W.G.: A methodology for selection problems with multiple, conflicting objectives and both qualitative and quantitative criteria. Int. J. Prod. Econ. **86**(3), 187–199 (2003)

15. Talluri, S., Baker, R.: A quantitative framework for designing efficient business process alliances. In: Proceedings of Engineering and Technology Management 1996 (IEMC 96), pp. 656–661 (1996)

16. O'Brien, C., Ghodsypour, S.H.: A decision support system for supplier selection using an integrated analytic hierarchy process and linear programming. Int. J. Prod. Econ. **56**(9), 199–212 (1998)

17. Hwang, C.L., Yoon, K.: Multiple Attribute Decision Making: Methods and Applications. Springer, New York (1981)

18. Suriadi, S., Weiß, B., Winkelmann, A., ter Hofstede, A.H., Adams, M.: Current research in risk-aware business process management — overview, comparison, and gap analysis. CAIS, vol. 34, pp. 933–984, Article 52 (2014)

19. Chinn, D.: What Is Business Process Management? (2012). http://www.ehow.com/info_7758354_business-process-management.html

20. Jakoubi, S., Tjoa, S., Goluch, G., Quirchmayr, G.: A survey of scientific approaches considering the integration of security and risk aspects into business process management. IEEE (2009). 1529-4188/09 – 2009

21. Polyantchikov, I., Shevtshenko, E., Karaulova, T., Kangilaski, T., Camarinha-Matos, L.M.: Virtual enterprise formation in the context of a sustainable partner network. Ind. Manage. Data Syst. **117**(7), 1446–1468 (2017)

22. Polyantchikov, I., Shevtshenko, E.: Partner selection criteria for virtual organization forming. In: Otto, T. (ed.) Proceedings of 9th International Conference of DAAAM Baltic Industrial Engineering, Tallinna Tehnikaülikooli Kirjastus, 24–26 April 2014, Tallinn, Estonia, pp. 163–168 (2014)

23. Polyantchikov, I., Bangalore Srinivasa, A., Veerana Naikod, G., Tara, T., Kangilaski, T., Shevtshenko, E.: Enterprise architecture management based framework for integration of SME to collaborative network. Collaborative Networks on the Internet of Services. In: 13th IFIP WG 5.5 Working Conference on Virtual Enterprises, PRO-VE 2012, Bournemouth, UK, October 2012

24. Polyantchikov, I., Karaulova, T., Shevtshenko, E., Kangilaski, T., Netribiitshuk, V.: Web environment elaboration working with ISO 9001 documents at a production enterprise. In: Zakis, J. (ed.) 13th International Symposium on Topical Problems in the Field of Electrical and Power Engineering, Doctoral Dchool of Energy and Geotechnology II, pp. 304–311 (2015)

25. Chen, C.-T.: A fuzzy approach to select the location of the distribution center. Fuzzy Sets Syst. **118**, 65–73 (2001)

26. Snatkin, A., Eiskop, T., Karjust, K., Majak, J.: Production monitoring system development and modification. Proce. Est. Acad. Sci. **64**, 567–580 (2015)

27. Karjust, K., Küttner, R., Pääsuke, K.: Adaptive web based quotation generic module for SME's. In: Küttner, R. (ed.) Proceedings of 7th International Conference of DAAAM Baltic Industrial Engineering, Tallinn, pp. 375–380, 22–24 April 2010. Tallinn University of Technology Press (2010)

28. Kangilaski, T.: Maturity models as tools for or focal player forming, virtual organizations. In: International Conference on Management, Manufacturing and Materials Engineering, Zhengzhou (2011)

29. Shevtshenko, E., Poljantchikov, I., Mahmood, K., Kangilaski, T., Norta, A.: Collaborative project management framework for partner network initiation. Procedia Eng. **100**, 159–168 (2015)

Reflections upon Measurement
and Uncertainty in Measurement

Carlo Ferri[(⊠)]

WMG, University of Warwick, Coventry CV4 7AL, UK
c.ferri@warwick.ac.uk

Abstract. Measurements of physical quantities are the corner stone upon which we humans have built the scientific perception of the world. They characterize the scientific system of beliefs: measurements are the distinctive means to tell the scientific truth apart from any other kind of approach to knowledge. Moreover, measurements have been having a central part in the development of any sort of machine, instrument and artefact that humans have invented. Yet in industry, especially in small medium size enterprises (SME's), time and money spent in measurements is often seen as a necessary evil or, worst, as a waste of valuable resources. To contribute in contrasting this negative perception, a clarification of the fundamental concept of measurement is presented. The emphasis is in particular placed on uncertainty in measurement. The need for the introduction of the concept of uncertainty is justified. The theoretical implications attached to uncertainty of measurement are analyzed.

Keywords: Measurement uncertainty
International Vocabulary of Metrology (VIM)
Guide to the Expression of Uncertainty in Measurement (GUM)

1 Introduction

Soft-computing may be described as the collection of those computational methods robust to indeterminacy (vagueness) and capable of giving likely-suboptimal yet sufficiently good solutions in a simple and reasonably time-inexpensive way [8]. For example, among these techniques there is fuzzy logic, a conceptual framework developed by Lofty Zadeh in 1965 within his theory of fuzzy sets. To account for vague, qualitative, indeterminate concepts, notions such as degree of membership of an element to a set and degree of truth of a proposition were introduced and developed in a consistent formal theory.

The fact that soft-computing methods have been specifically introduced to overcome the difficulties in handling vagueness and qualitative knowledge in computational environments has generated a quite widespread misconception: once a quantity has been measured, then all the vagueness has vanished, because the quantity is described by a

---

Part of this investigation was performed when the author was at Coventry University, Faculty of Engineering and Computing, Gulson road, Coventry CV1 2JH, United Kingdom.

© Springer Nature Switzerland AG 2019
B. K. Ane et al. (Eds.): WSC 2014, AISC 864, pp. 234–243, 2019.
https://doi.org/10.1007/978-3-030-00612-9_20

number. In truth, instead, any measurement result is inherently uncertain. In this sense, therefore measurements pertain in their own right to the domain of knowledge where soft-computing techniques may be applied. In this piece of writing frequent reference is made to two of the most authoritative sources of reference in metrology, namely: the International Vocabulary of Metrology (VIM) [10] and the SI Brochure [2]. In the next section the concept of measurement is explored. In Sect. 3 the indeterminacy inherent into the comparison of a quantity with a unit of measurement is presented. The impossibility of obtaining certainty in the realization of unit of measurements is highlighted in Sect. 4. The impossibility of defining completely and unambiguously any measurand is then described in Sect. 5. A discussion follows and conclusions are drawn thereafter.

2 Measurements

Measurement is any experimental process aimed at obtaining one or more numbers and a reference that can be attributed to a property of a body, a phenomenon or a substance (cf Sects. 1.1, 1.19, 2.1 in [10]). This property is called a quantity and its magnitude is defined as the number and the reference considered together. The reference typically is a measurement unit (e.g. the kilogram, when measuring a mass), but it can be a measurement procedure (e.g. Rockwell C, when measuring hardness) or a reference material (e.g. the concentration of luteinizing hormone in a specimen of human blood plasma, cf Sects. 1.1 and 1.19 in [10]). The Measurement unit is a quantity selected conventionally to which any other quantity of the same kind can be compared. The result of this comparison is called the ratio of the two quantities and is expressed as a number (cf Sect. 1.9 in [10]).

From a logic perspective, it then follows that three conditions are necessary for a measurement result not to be intrinsically uncertain:

(a) It should always be possible to compare the measurand (i.e. the quantity intended to be measured, cf Sect. 2.3 in [10]) and the measurement unit so that no indeterminacy is present in the numerical quantity value (cf Sect. 1.20 in [10]);
(b) The unit of measurement should have an unambiguous magnitude;
(c) The measurand should be defined without any indeterminacy.

Unfortunately, none of these conditions holds, as it is described in Sects. 3, 4 and 5 for (a), (b) and (c), respectively.

3 Comparisons of the Measurand to the Unit

The comparison of the measurand and the measurement unit is achieved by the interaction of the body, phenomenon or substance under study and a measuring system that produces an indication sensitive to the measurand. A measuring system is any set of devices that is designed to generate measured quantity values (cf Sects. 3.2 and 2.10 in [10]). Typically, the nature of the interaction measurand-measuring system cannot be isolated from other quantities characterising the conditions in which the measurement takes place. For example, when measuring the height of a table with a tape measure the measured height value is not only a function of the height of the table, but also of a number of other quantities. Among these, there are for instance the field of air temperature and air humidity affecting the wood of the table, the temperature of the hands of the person holding the tape measure, the resolution of the tape measure, the discretion of the person reading the indication of the tape measure when judging the alignment of the tape with the table and the alignment of the scale on the tape with the extremes delimiting the table height. This interdependence between quantities is ideally captured by 'a mathematical relation among all quantities known to be involved in a measurement' which is called the measurement model (cf Sect. 2.48 in [10]). Namely, it holds:

$$h(Y, X_1, X_2, \ldots, X_n) = 0 \tag{1}$$

In Eq. 1, Y is the measurand or output quantity of the model whereas the other quantities X_1, X_2, ..., X_n are called the input quantities in the measurement model. The value of the output Y is to be calculated from that of the input quantities. Often Eq. 1 can be explicitly defined as follows:

$$Y = f(X_1, X_2, \ldots, X_n) \tag{2}$$

In Eq. 2, the function $f(X_1, X_2, \ldots, X_n)$ is referred to as measurement function (cf Sect. 2.49 in [10]). This situation generates indeterminacy of the measurement in at least two different ways. First, the input variables in a measurement are not uniquely

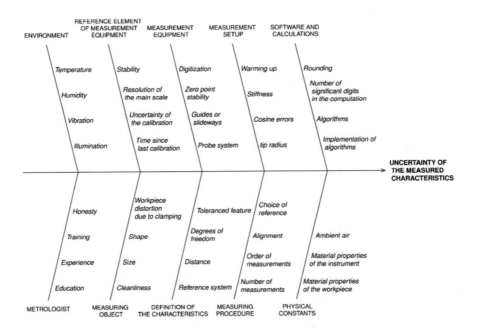

Fig. 1. An Ishikawa (fishbone) diagram with a few potential sources of uncertainty that are typically investigated for dimensional measurement results. The sources are grouped according to ISO 14253-2:2011 [1]

known. The input variables included in a model depend on the expected use of the measurement result. For example, a measurement result having implications on the life of many would justify the investigation of a large number of potential input variables for inclusion in the measurement model. The extra cost incurred for the broad array of instruments needed to measure these input variables would be justified. Second, the measuring model $h(\dots)$ (or function $f(\dots)$) is typically not known analytically, barring the cases when some physical model of the measurement is available. Hence the effect of an input variable upon the output must be estimated. Figure 1 is inspired to Fig. 4 in ISO 14253-2:2011 [1]. It displays a grouping of candidate input variables. The figure suggests a possible systematic procedure for selecting input variables for analysis without omitting some relevant group of them. In this context, the term uncertainty used in the figure can be interpreted in its generic meaning of vagueness or indeterminacy. The figure has been produced using a free software package [12] (free as freedom not as gratis).

4   Units

The units of measurement do not have an unambiguous unique magnitude. To support this statement, the concept of 'definition of a unit' must be distinguished and kept apart from the concept of 'realization of a unit', as explained in the SI Brochure (cf page 111, Sect. 2.1.1 in [2]). A measurement unit is a quantity that is conventionally defined so as it has solid theoretical foundations and it enables measurements as reproducible as possible. The realization of a unit is instead a procedure where a quantity value is associated to a quantity of the same kind as the unit and that fulfils the definition given for the unit. In other words, the realization of a unit is a measurement assigning a magnitude to a realized unit, i.e. to a quantity existing in the sensory world and not just on the paper as in the unit definition. The indeterminacy and vagueness discussed above for the concept of measurement does then generate the indeterminacy in the magnitude of the realized unit. For example, the unit of length in the International System of Units (SI – Le Système International d'Unitès) is the metre which is defined in the SI Brochure as 'the length of the path travelled by light in vacuum during a time interval of 1/299 792 458 of a second' (cf page 112, Sect. 2.1.1.1 in [2]). Guidelines for the realization of the metre are instead presented in what are referred to as the *mises en pratique*, which are published on the web to facilitate frequent revision [3, 4].

5   Measurands

To define unambiguously a measurand an infinite amount of information is needed. For example, to define the height of a table, a point P_1 on the top surface of the table could be identified. A straight line l_1 orthogonal to that surface and passing through P_1 could be constructed. Let P_2 be the point of intersection between l_1 and a second surface representing the floor. The height of the table can be defined as the distance between P_1 and P_2. The example is illustrated in Fig. 2 where two different distances satisfying the measurand definition above are shown (l_1 (A) and l_1 (B)). In fact, in this definition of height of a table, there are infinite possible choices for the starting point on the top surface. The height of the table is deeply affected by such a choice. In Fig. 3, two measurement results of the two distances satisfying the definition of height of a table as proposed above are displayed. The measurement results have been obtained using an articulated arm coordinate measurement machine driven by a proprietary software widely used in industry. Moreover, a large variety of different types of surfaces can be selected to represent the top surface of the table and the floor: from the natural choice of a plane shown above, to that of some more complex non-uniform rational basis spline surfaces (NURBS surfaces). In addition, a range of different choices can be made when associating the chosen type(s) of surface, which is an abstract entity of the human rationality, to a physical table or floor, which are sensory entities perceived by humans using their senses (sight and touch in this case). This unavoidable intrinsic vagueness in the definition of a measurand is called definitional uncertainty (cf Sect. 2.27 in [10]).

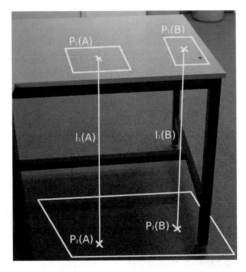

Fig. 2. Two different distances both compliant with the definition of the measurand height of a table introduced above

6 Discussion

The argumentation presented so far is aimed to make the intrinsic indeterminacy of measurement results apparent. In the Guide to the Expression of Uncertainty in Measurement (GUM), this indeterminacy or vagueness that expresses a doubt about the result of a measurement is referred to as uncertainty (cf Sect. 2.2.1 in [9]). In the same document, the term uncertainty is however also used in a more specific way to designate a parameter providing a quantitative measurement of this generic concept of doubt. Namely, in the GUM uncertainty is defined as a 'parameter, associated with the result of a measurement, that characterizes the dispersion of the values that could

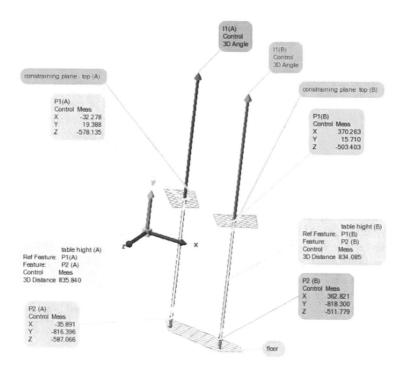

Fig. 3. Two measurement results of the two heights of a table displayed in Fig. 2

reasonably be attributed to the measurand' (cf Sect. 2.2.3 in [9]). In the VIM instead, measurement uncertainty is defined as a 'non-negative parameter characterizing the dispersion of the quantity values being attributed to a measurand, based on the information used' (cf Sect. 2.26 in [10]). Typically, this parameter is the standard deviation of a probability density function that models the incomplete or partial state of knowledge of the measurand achievable with measurements.

Acknowledging that measurements are always unavoidably uncertain has some profound implications on how humans construct their knowledge about the physical world in science and technology. Measurements are the ultimate source of knowledge in science: any statement to be scientifically accepted must be substantiate by experiments or observations that are expressed in terms of measurement results. If measurement results are inherently uncertain, all what can be inferred from them can be only uncertain. In other words, talking of 'exact science' when referring to Physics, for example, can be quite prone to misinterpretations. Science may be considered exact only in its methods of dealing with approximations and uncertain or partial knowledge. Stretching this view to its extreme may lead to consider science as an activity with very useful practical effects but with little use in the unambiguous identification of the truth.

Recognizing that measurements are uncertain is a fact that humans can exploit in acquiring new knowledge about the physical world. Investigating the uncertainty structure in designed experimental conditions may enable experimenters to add new contributions to knowledge. Examples of how the characterization of uncertainty fosters the acquisition of new knowledge have been directly experienced by the author in the investigation two different machining processes (micro electric discharge machining and contamination-free turning [6, 7]).

The practical importance of uncertainty in measurement is perhaps epitomized by all those cases where decisions have to be made on the basis of measurement results. For example, how could a plaster cast of a sculpture be distinguished from the original work of art on the basis of its form alone? In Fig. 4 the digitization process and the resulting digital model of a 19th century plaster cast of the head of David by Michelangelo is displayed. The original David is a masterpiece of Renaissance sculpture created by Michelangelo between 1501 and 1504 currently housed in the Gallery of the Academy of Florence (Firenze, Italy). Even if the plaster cast were identical to the original, the uncertainty in measurement inherent in the digitization process would prevent the virtual model to be identical to the plaster cast. Likewise, if a digital model of the original David were created in the same way, also that digital model would be affected by uncertainty in measurement. Then a comparison between the two digital models can only happen in probabilistic terms. Likewise, any statement regarding the identity between the plaster cast and the original sculpture made on the basis of these measurement results would necessary be of a probabilistic nature.

Fig. 4. Scanning of a 19th century plaster cast of David by Michelangelo (left). The plaster cast is housed in the 'Michelangiolesco Museum' of Caprese Michelangelo (Italy). The digital model (right) is made of 37,035,104 points (courtesy of cam2 *s.r.l.*, Grugliasco, Italy)

Soft-computing methods with their ability to account for indeterminate and partial knowledge may provide alternative approaches to account for uncertainty in measurements. The envisaged possible benefit is that these alternative perspectives may in turn promote and facilitate the acquisition of new knowledge in science and technology. From an industrial point of view, a number of software applications have been developed and commercialized to assist practitioners in the evaluation of the uncertainty in their measurements. In Fig. 5 is displayed the graphical output of a Monte Carlo simulation of the measurement of coordinates of a field of constructed points in space. The measurements being simulated refer to a laser tracker that has been set up in two different locations. In the simulation, each point has been measured 1000 time from each of the two tracker positions. In the figure, the two sets of 1000 measured points for each constructed nominal point are clearly different for the two tracker positions. Each of these sets of 1000 points is sometime referred to as point uncertainty field. The simulation allows a practitioner to assess visually the effect that the location of the instrument has on the reliability of the measurement taken. This kind of simulations are quite widespread in industrial portable metrology, most typically in the aerospace industry.

Fig. 5. Graphical representations of Monte Carlo simulation of point coordinate measurements taken with the same laser trackers placed in two different positions (starting from above: top view and axonometric projection, respectively)

From a research perspective, the main issue with commercially-available software applications as the application used for Fig. 5 is that the large majority of the software houses active in this field operate a business model centered around restrictive licenses. They tend not to make the information regarding the models and algorithms available to their customer base. To contrast this difficulty, some researchers have endeavored to build their own tools making them available under one of the licenses from the Free Software Foundation. For example, within the free software environment for statistical computing and graphics called R [11], a couple of packages regarding measurements and uncertainty are *MetRology* [5] and *propagate* [13]. They enable the users to make Monte Carlo and Bayesian evaluations of uncertainty within R. Some consideration of the benefits and limitations of applying soft-computing techniques in measurements problems and in uncertainty evaluations in particular has been given in the past [14]. However, the relationship between soft-computing and metrology still appears a promising open field of investigation.

7 Conclusions

An analysis of the concept of measurement and measurement result has been presented. This analysis supports the idea that even when a quantity is measured, it is not completely known. This partial state of knowledge has been ascribed to three sources of indeterminacy unavoidably-attached to any measurement result: the measurement process itself, the unit of measurement and the definition of the measurand. Uncertainty as a technical term introduced by authoritative international bodies as a means of representing this intrinsic indeterminacy in measurements has been discussed. Soft-computing methods may offer alternative models of measurement uncertainty facilitating the acquisition of new knowledge of the physical world we all live in.

References

1. BS EN ISO 14253-2:2011 geometrical product specifications (GPS) inspection by measurement of workpieces and measuring equipment. Part 2: Guidance for the estimation of uncertainty in GPS measurement, in calibration of measuring equipment and in product verification (2011)
2. Le système international d'unités (SI) - The international system of units (SI). Bureau International des Poids et Mesures (BIPM) - Organisation Intergovernementale de la Convention du Mètre. visited in December 2014, 8th edition, ISBN 92-822-2213-6 (2006). http://www.bipm.org/en/publications/si-brochure/
3. Mises en pratique - practical realizations of the definitions of some important units. http://www.bipm.org/en/publications/mises-en-pratique/. Accessed in December 2014 (2014)
4. Mises en pratique - practical realizations of the definitions of some important units: Recommended values of standard frequencies. http://www.bipm.org/en/publications/mises-en-pratique/standard-frequencies.html. Accessed in December 2014 (2014)
5. Ellison., S.L.R.: metRology: Support for metrological applications (2014). http://CRAN.R-project.org/package=metRology. r package version 0.9-17

6. Ferri, C., Ivanov, A., Petrelli, A.: Electrical measurements in μ-edm. J. Micromech. Microeng. **18**(8), 085007 (2008). http://stacks.iop.org/0960-1317/18/i=8/a=085007

7. Ferri, C., Minton, T., Ghani, S.B.C., Cheng, K.: Efficiency in contamination-free machining using microfluidic structures. CIRP J. Manuf. Sci. Technol. **7**(2), 97–105 (2014). http://www.sciencedirect.com/science/article/pii/S1755581713000746

8. Hajek, P.: Fuzzy logic. In: Zalta, E.N. (ed.) The Stanford Encyclopedia of Philosophy. Fall 2010 edn. (2010)

9. JCGM - Joint Committee for Guides in Metrology: JCGM 100:2008. Evaluation of Measurement Data – Guide to the Expression of Uncertainty in Measurement, first, corrected version 2010 edn. (2008)

10. JCGM - Joint Committee for Guides in Metrology: JCGM 200:2012. International vocabulary of metrology – Basic and general concepts and associated terms (VIM), third, version with minor corrections edn. (2008)

11. R Core Team: R: A Language and Environment for Statistical Computing. R Foundation for Statistical Computing, Vienna, Austria (2014). http://www.R-project.org/

12. Scrucca, L.: qcc: an R package for quality control charting and statistical process control. R News **4**(1), 11–17 (2004). http://CRAN.R-project.org/doc/Rnews/

13. Spiess, A.N.: Propagate: Propagation of Uncertainty, r package version 1.0-4 (2014). http://CRAN.R-project.org/package=propagate

14. Varkonyi-Koczy, A., Dobrowiecki, T., Peceli, G.: Measurement uncertainty: a soft computing approach. In: IEEE International Conference on Intelligent Engineering Systems, Proceedings, INES 1997, pp. 485–490, September 1997

Author Index

© Springer Nature Switzerland AG 2019
B. K. Ane et al. (Eds.): WSC 2014, AISC 864, p. 245, 2019.
https://doi.org/10.1007/978-3-030-00612-9

Printed in the United States
By Bookmasters